16·50

The Science and Technology of Foods

R K Proudlove
BSc MPhil FIFST

Principal Lecturer, External Liaison
Lincoln University
School of Applied Science & Technology

FORBES   PUBLICATIONS

The molecular model in our cover
photograph is a dipeptide of Methionine
and Phenylalanine, and was constructed
from a kit kindly loaned by Griffin &
George. Glassware, cutlery and crockery
were loaned by David Mellor Ltd.
Photograph by Steve Marwood.

All rights reserved
First published 1985
Second edition 1989
Third edition 1992
Reprinted 1994
Reprinted 1997
ISBN 0 901762 90 3

Printed in Great Britain by
St Edmundsbury Press Limited,
Bury St Edmunds, Suffolk

Contents

Preface

During the last few years there has been a considerable increase in the Food Science and Technology taught in 'A' level courses in Home Economics, and to a lesser extent in Social Biology. A similar trend is now emerging at GCSE level. This book is planned to meet the needs of these syllabuses in respect of their Food Science and Technology content. The relevant aspects of nutrition are also included, but the work is not intended to be an exhaustive study of nutrition.

Many aspects of the study of foods have been broadened and deepened to meet the needs of college courses at Diploma and Higher Diploma level in Home Economics, Catering, and Food Technology. The book will also serve as a general background for the first year of some 'food' degree courses.

The study of food is an endless subject and each topic area is sufficient to fill a book in its own right. In order to cover a wide range of topics the book is divided into four sections.

Section 1, The Composition of Foods, is devoted to the study of individual food components. Certain aspects of chemistry cannot be avoided, but these have been treated as simply as possible. In *Section 2,* the work of Section 1 is applied to the study of commodities and raw materials. A study of each food is undertaken, together with relevant aspects of handling and processing. The emphasis changes in *Section 3* to a description of those processes which occur in the handling and preparation of food raw materials. The production of convenience foods is the theme of the section. *Section 4* is devoted to those processes which preserve the food, and also to packaging technology.

After each topic area, a small review section, in note form, is included to help the student in revision and to help the teacher to scan quickly the contents of the chapter. Wherever possible practical exercises have been included to illustrate a particular principle. The experiments have been planned to illustrate food processing operations wherever possible, and have been designed around the minimum of equipment and materials.

R.K.P.

Introduction to the Third Edition

Although the food processing industry has always been aware of maintaining quality of food there is now a greater emphasis placed on the industry on safety and quality. The Food Safety Act of 1990 has widened this concern for safety from the 'farm to the table'. A number of regulations emanating from the Act are now being put into effect. In particular the Food Act places great emphasis on temperature control and monitoring. Companies will have to install refrigeration equipment or upgrade existing equipment to maintain low temperatures. The companies will also have to maintain records proving that the required temperatures have been achieved.

The new temperature control regulations came into effect on 1 April 1991. Food which is sold or served hot must be at a temperature of at least 63°C or otherwise chilled. Chilled temperatures must be below 8°C for most cooked, smoked, cured fish and meat product. By April 1993 temperatures are to be reduced to 5°C. In practice many companies and supermarkets now store below 4°C and often close to 0°C, but avoiding any freezing.

Further significant regulations came into effect on January 1 1991 and this was to legalise the irradiation of some food products including some vegetables, spices and poultry.

The controversy surrounding irradiation is still going on and consumer resistance still seems to be as great. So it is still unlikely that we will see irradiated food in the shops in the near future.

Increasing numbers of food companies and their suppliers are applying for British Standard registration, through the standard BS5750 (international equivalent EN29000/1S0 9000). Very high standards of quality have to be met so that registered companies will supply goods of a consistently high quality. Quality thus becomes the responsibility of everyone working within the company. There is, therefore, a growing trend towards total quality management (TQM), with the operatives within the factories having a key role to play in assuring product quality.

Needless-to-say there is still much activity in the field of nutrition. In 1991 the COMA committee produced a report updating its previous findings. Again there were recommendations for reductions in the nation's intake of salt, fat and sugar and an increase in the consumption of fibre and vitamin C. The food industry has been pressurised to provide more low-fat, low sugar products. Unless there is a surge in consumer demand the industry is reluctant to change whole product lines, which are profitable, to meet these requirements, particularly as alternatives are readily available.

The COMA report introduced a new concept of Dietary Reference Values (DRVs) for 33 nutrients, 13 vitamins and 15 minerals known to be important for human health. These values are said to be more accurate than Recommended Daily Amounts (RDAs) of food energy and nutrients which are used as the current standard. DRVs are an estimate of the quantity of each nutrient required by a person to achieve a balanced diet. The DRVs for sugars, fats and starches are given as a percentage contribution to total food energy and other nutrients as grams per day.

The new guidelines are similar to previous ones emanating from NACNE and similar reports. However, fat consumption should be reduced from 42% of total food energy to 35%, and saturated fat reduced from 16% to 11%. Sugars should be slightly reduced from 13% to 11%. Salt is still in dispute but generally it is agreed a reduction from 8 g per day on average to 4 g would be beneficial. Starch consumption should be increased from 28% to 39% of total food energy. Vitamin C consumption should be increased to 40 mg from an average of 30 mg per day. Many people still do not have a sufficient fibre intake in their diet and it is recommended that dietary fibre should be increased from 12 g to 18 g a day.

Unfortunately, as market research has shown, perhaps as many as 50% of consumers do not care about their intake of sugar, salt or caffeine.

The use of food additives has declined as a result of consumer and media pressure. However, there is a growth in the use of specialised ingredients which give consumers some nutritional or even therapeutic benefit, and these may well play a significant role in the food industry during the 1990's. These functional foods have been termed 'nutriceuticals' and they contain naturally occuring components with specific functions. For example, oat bran-based products have been linked with a reduction in serum cholesterol. Similarly, omega-3 (N-3 fatty acids from oily fish have been shown to lower the risk of cardiovascular disease. This has led to the development of omega 3 based margarines.

The 1990s will see rapid development in food processing, handling and in new products. There is a need to feed a world population which is starving in many places. There is enough food to do this now but better means of handling and distribution are needed.

R. K. Proudlove

Acknowledgement

I would like to thank Miss M. Phillingham,
Education Inspector for Home Economics
(Lincolnshire), for advice and helpful comments in
preparing this book.

Section 1

The Composition of Foods

The majority of foods are complex mixtures of a large number of different compounds. However, these mixtures are rarely constant. The chemical composition of a fruit, for example, will vary according to variety, stage of ripeness, climate, growing conditions, position on the tree and even in some cases atmospheric pressure. Similarly, the composition of cow's milk varies according to breed, stage of lactation, diet and time of year. Figures quoted for levels of a component in a food are often only average figures and an analysis of the food may give a somewhat different figure.

The general composition of foods is summarized in Figure 1.1, and, with the exception of dried and concentrated foods, water is always the main component, often 80–90% by weight of the food.

Figure 1.1 The composition of foods

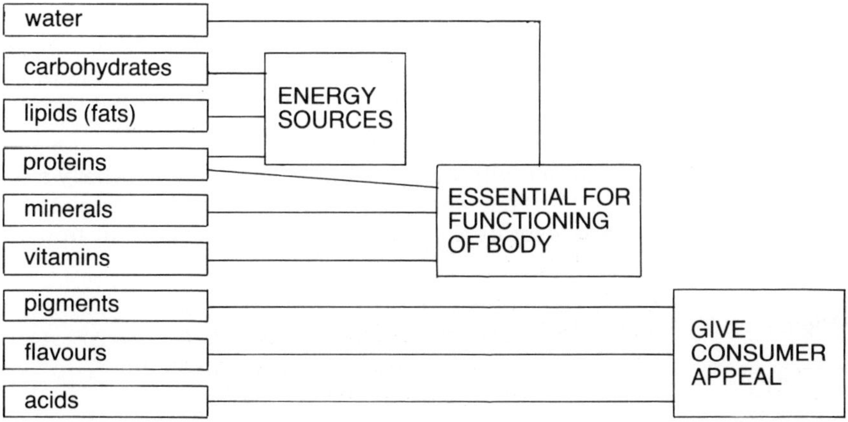

In addition to the components listed in Figure 1.1, most manufactured or processed foods contain a number of additives to fulfil various functions.

A particular food component, itself, may be made up of a large number of smaller components; for example a fruit flavour may have over 100 different constituents.

1

1.1 Water

Although water is the main component of most foods, surprisingly many other food components, eg fats, proteins, some vitamins and pigments are unable to dissolve in it. As well as being essential to life, water plays a very important role in the behaviour of foods, particularly during cooking and processing.

In order to understand why water is important in food in this way let us first consider its structure. The water molecule contains two atoms of hydrogen bound to one atom of oxygen by the sharing of electrons.

However, the oxygen atom has the ability to attract electrons towards itself and so the sharing of the electrons is uneven between the oxygen and hydrogen atoms. This has the effect of giving the oxygen atom a partial negative charge ($\delta-$) and the hydrogen atoms a corresponding positive charge ($\delta+$) to balance this.

Figure 1.2 Structure of a water molecule

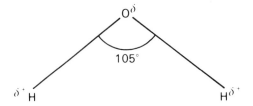

(*NB* You may find in some books that O is given $2\delta^-$ or H a $\frac{1}{2}\delta^+$ to show a balance between the charges. However, δ means a partial or a 'bit of' charge and therefore it is preferable to leave it simply as given in the figure.)

A 'North Pole' of a magnet will attract a 'South Pole' and similarly a δ^- charge will attract a δ^+. This weak electrostatic attraction is called a *hydrogen bond*, and is usually represented by three dashes in line, as shown in Figure 1.3.

Figure 1.3 Hydrogen bond

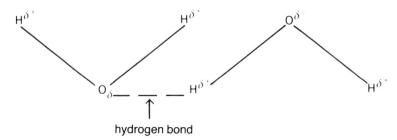

hydrogen bond

In water, hydrogen bonds are constantly being formed and then broken. If the temperature rises molecular movement increases and thus hydrogen bonds decrease. Conversely, as the temperature falls water molecules move less so more hydrogen bonds form until finally so many exist that water solidifies into ice.

Hydrogen bonding occurs widely in foods and is responsible for some very important phenomena. For example, hydrogen bonds are involved in the gelatinization of starch, the setting of jams, and the tenderness of steak. Although these bonds are very weak they often occur in considerable numbers and therefore can have a combined effect in a food. A common example found in many food products is the linking together of long polymer chains such as polysaccharides. The hydrogen bonding may link directly from one chain to the next by means of hydroxyl groups (OH) or may involve a water molecule as a bridge.

Figure 1.4 Linking of polymers by hydrogen bonds

(i)

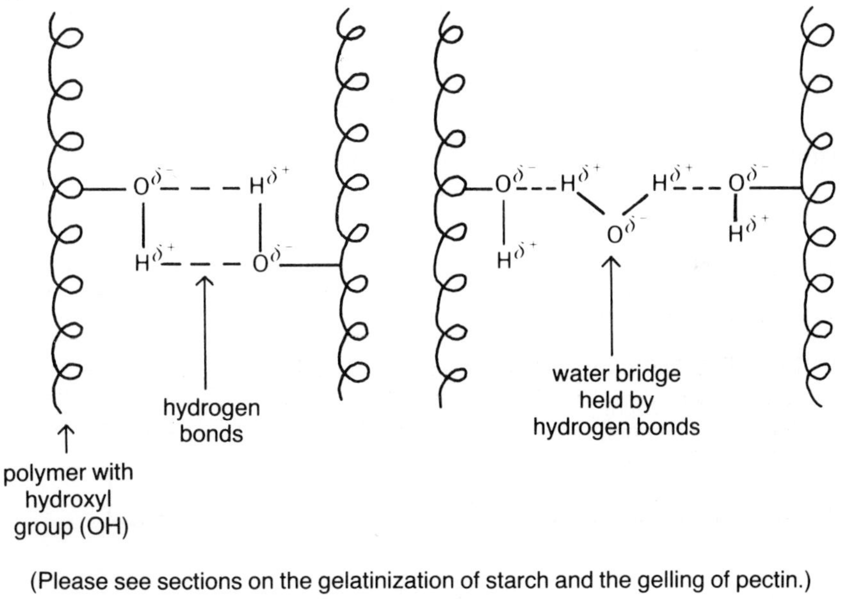

polymer with
  hydroxyl
  group (OH)

hydrogen
  bonds

water bridge
held by
hydrogen bonds

(Please see sections on the gelatinization of starch and the gelling of pectin.)

1.2 Carbohydrates

Over half the organic matter on the earth is carbohydrate in one form or another. The most widespread is cellulose. Although cellulose is not digested by humans it is nevertheless important in the diet as a main contributor of dietary fibre. Starch is the most common carbohydrate in human food.

The process of photosynthesis is responsible for producing carbohydrates in plants and is generally represented by a simplified equation:

$$6CO_2 + 12H_2O \xrightarrow[\text{chlorophyll}]{\text{sunlight}} C_6H_{12}O_6 + 6O_2 + 6H_2O$$

In reality a complex series of chemical changes occurs and may continue to produce a whole range of carbohydrates.

In general carbohydrates may be represented as: $C_x(H_2O)_y$. Hence, the name 'hydrates of carbon' or carbohydrates.

There are a number of different types of carbohydrates and the most common ones found in food are given in Table 1.1.

Table 1.1 Classification of commonly occurring carbohydrates

Sugars (sweet)	Monosaccharides	Glucose Fructose
	Disaccharides	Maltose Sucrose Lactose
Non-sugars (not sweet)	Simple Polysaccharides	Starch Cellulose Glycogen
	Complex Polysaccharides	Pectin Gums

1.2.1 Sugars

1.2.1.1 Monosaccharides

These sugars contain from 2 to 7 carbon atoms, but the most common ones in foods contain 6 (hexoses) and occasionally 5 carbon atoms (pentoses). The formula: $C_6H_{12}O_6$ refers to *any* monosaccharide with 6 carbon atoms and not just to the most common monosaccharide, glucose.

Glucose

Glucose or dextrose occurs widely in fruits, onions, potatoes and is used in many manufactured foods.

As pointed out above, like all hexoses, the formula of glucose is $C_6H_{12}O_6$ but this conveys very little. Written as a structural formula it can be represented as shown in Figure 1.5.

Figure 1.5 Structural formula of glucose

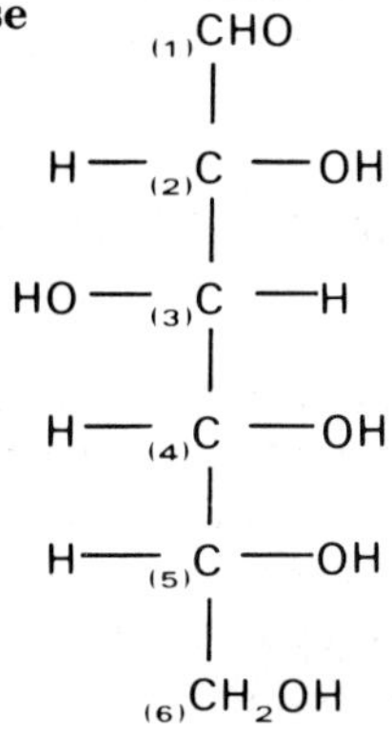

((1)–(6), carbon atoms)

In reality this straight chain form does not exist, but glucose normally takes the form of a six-sided structure.

Figure 1.6 α-glucose and β-glucose

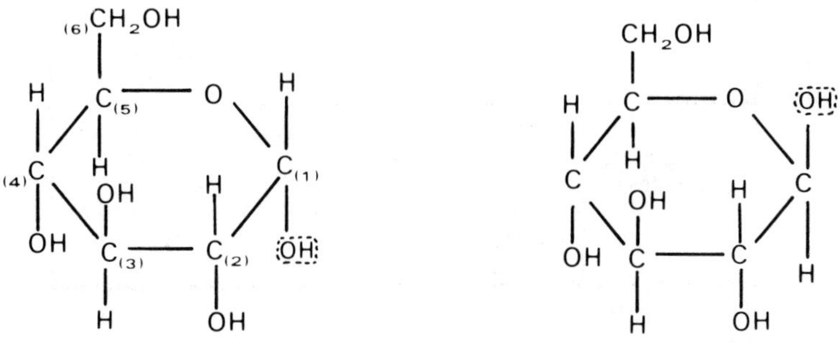

α-glucose

β-glucose

(NB When writing these formulae always check that they add up to $C_6H_{12}O_6$.)

There are two forms of glucose shown above: α-glucose with the hydroxyl group (OH) on carbon atom 1 at the bottom, β-glucose with the hydroxyl group at the top. It is important to remember that this is the only difference between the two glucoses shown.

In the ring structure the six-sided figure is represented as a plane at right angles to the paper and the groups of −H and −OH are above or below this plane. It is conventional to miss out the carbon atoms in such a structure and a bend in the ring indicates the presence of a carbon atom as shown in Figure 1.7.

Figure 1.7 Ring structure of glucose

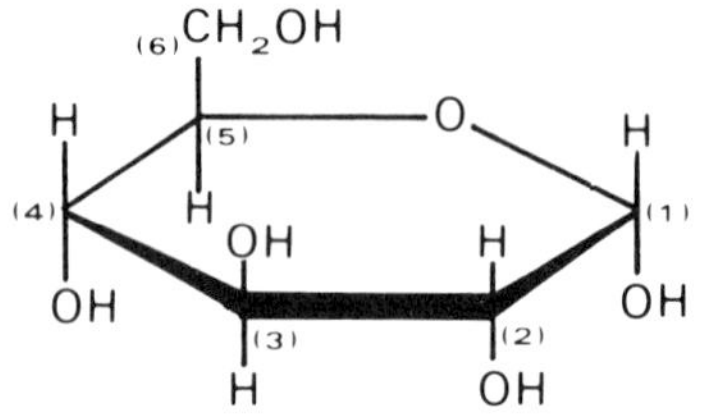

(1)–(6) = carbon atoms

(NB The thick part of the ring is effectively the closest part of the plane to you. This ring structure is called the configuration of glucose.)

More recent work has shown that the plane is not flat but has a certain shape and in most cases this shape is said to resemble a 'chair'.

Glucose is a *reducing sugar*. This means it has the ability to breakdown Fehling's solution to form a brick red coloured precipitate.

Fehling's solution is, in fact, a mixture of two solutions which have to be mixed at the time of carrying out the test for a reducing sugar.

Fehling's 1 (or A) –   Copper sulphate solution
Fehling's 2 (or B) –   A mixture in solution of sodium hydroxide and
                                   Rochelle salt (sodium potassium tartrate)

After mixing equal amounts of the two solutions the sugar is added and boiled. A brick red precipitate of copper I oxide (cuprous oxide) is formed if the sugar is a reducing sugar.

'*Glucose syrup*' is often used in the food industry, particularly in confectionery manufacture. It is never pure glucose but a mixture of glucose, maltose and other carbohydrates of higher molecular weight called dextrins. The term *dextrose equivalent* (DE) is used to indicate the level of glucose and maltose in the syrup. A low DE syrup is less sweet than a high DE syrup, as the latter contains more glucose.

Fructose

Fructose or laevulose is about one and half times sweeter than glucose, but it is often found with glucose in many foods, particularly fruits. It is a reducing sugar and because of its sweetness it is of importance in confectionery manufacture. Both fructose and glucose show *optical activity* in solution. If a plane of polarised light, i.e. light with wave motion in one plane only, is passed through a solution of the sugar, the plane is rotated in one direction. This can be seen using a polarimeter.

Figure 1.8 Optical activity of sugars

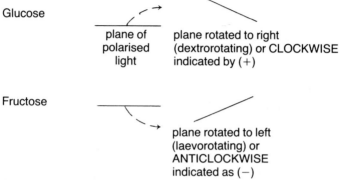

Glucose

plane of polarised light

plane rotated to right (dextrorotating) or CLOCKWISE indicated by (+)

Fructose

plane rotated to left (laevorotating) or ANTICLOCKWISE indicated as (−)

As glucose is dextrorotatory it has the alternate name of dextrose, and similarly the laevorotatory fructose can be called laevulose.

A mixture of equal amounts of glucose and fructose is known as *invert sugar*. Invert sugar occurs naturally in honey and is produced in jams during the boiling of the fruit with sugar. Invert sugar is always produced by the splitting of sucrose (sugar) into equal amounts of fructose and glucose. Bees produce the invert sugar in honey when collecting nectar, which is mainly sucrose, by the action of enzymes in their bodies. Acid in fruit will similarly split sucrose, and the process is called *inversion*.

Figure 1.9 Inversion of sucrose

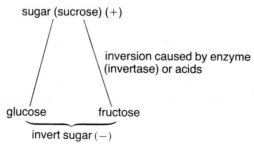

sugar (sucrose) (+)

inversion caused by enzyme (invertase) or acids

glucose   fructose

invert sugar (−)

Sugar in solution is dextrorotatory (+). Fructose is more strongly laevorotatory than glucose is dextrorotatory, and therefore sucrose (+)

7

is converted or 'inverted' to (−) invert sugar. Hence the term *inversion* is related to the change from (+) to (−) in optical rotation.

Fructose has a difficult structure to remember and this is further complicated by the fact that in the free state fructose is similar to glucose in having a six-sided structure. However, when combined with other sugars, for example with glucose to form sucrose, fructose exists as a five-sided structure.

Figure 1.10 Structure of fructose

fructose
(in free state)

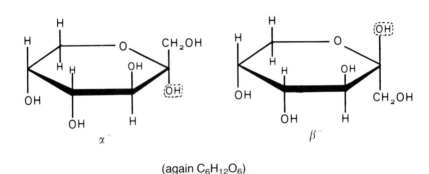

(again $C_6H_{12}O_6$)

fructose
(when combined
with other sugars)

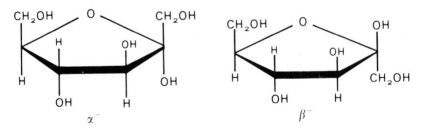

(again $C_6H_{12}O_6$)

1.2.1.2 Disaccharides

Disaccharides are formed from two monosaccharides which may be the same or different. In nature this combination is easily carried out under the control of enzymes, but it is impossible to achieve in the test-tube. Two monosaccharides condense together and eliminate water:

$$C_6H_{12}O_6 + C_6H_{12}O_6 \longrightarrow C_{12}H_{22}O_{11} + H_2O$$

Thus, the general formula for all disaccharides, ie disaccharides made up of two hexoses (six carbon atoms), is $C_{12}H_{22}O_{11}$. In foods, of the many known disaccharides only three are of importance. These are maltose, lactose and sucrose.

Maltose

Maltose or malt sugar is a simple disaccharide made from two glucose units. The two glucose units are linked across from carbon atom (1) on the left-hand glucose to carbon atom (4) on the right-hand glucose.

Figure 1.11 Structure of maltose

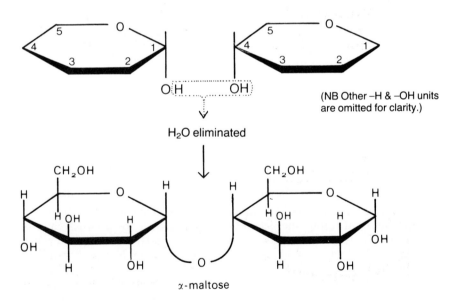

(NB Other –H & –OH units are omitted for clarity.)

H_2O eliminated

α-maltose

(NB A β-maltose does exist but is less common; its structure is the same except the OH on the carbon atom 1 on the right-hand side is at the top.)

When the two glucoses condense together water is eliminated and the remaining oxygen atom forms a bridge between the two glucoses. This bridge is called a *glycosidic link*. In this case the glycosidic link is called an α, 1–4 link, because the left-hand sugar is an α-form and the link is between carbon atoms 1 and 4 of the two sugars joined.

Figure 1.12 α,1–4 glycosidic link

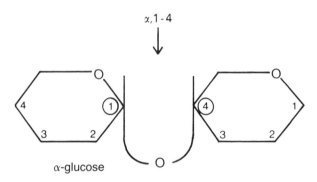

Maltose is a reducing sugar and is produced when starch is broken down by the action of enzymes (amylases), particularly in the malting process of barley for beer making. (See section on starch, page 13).

Lactose

This disaccharide is only found in milk. Lactose occurs in different amounts in cows' milk (4–5%), compared with human milk (6–8%). This fact has to be taken into account when manufacturing baby food from cows' milk. Lactose, in its pure state, is a white crystalline solid. Crystals of lactose can sometimes be found in cans of sweetened condensed milk.

After concentrating milk by evaporating some of the water it contains, a large amount of sugar (sucrose) is added to preserve the product. Sucrose is more soluble than lactose in water, and as there is limited water available, the lactose is forced out of solution. If the lactose crystallises slowly it may form crystals as big as marbles. However, it is usual to seed the sweetened condensed milk with very small crystals of lactose which causes the lactose to crystallize out quickly and thus only form very small crystals which can hardly be noticed.

Lactose is a reducing sugar and is composed of the two monosaccharides, galactose and glucose. The galactose is in the β-form combined with α-glucose to form α-lactose or with β-glucose to form β-lactose.

Figure 1.13 Structure of lactose

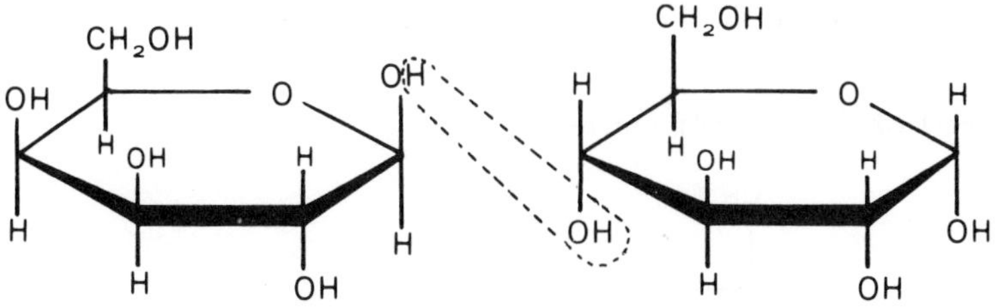

β-galactose α-glucose

(note the only difference with
β-glucose is on carbon atom 4)

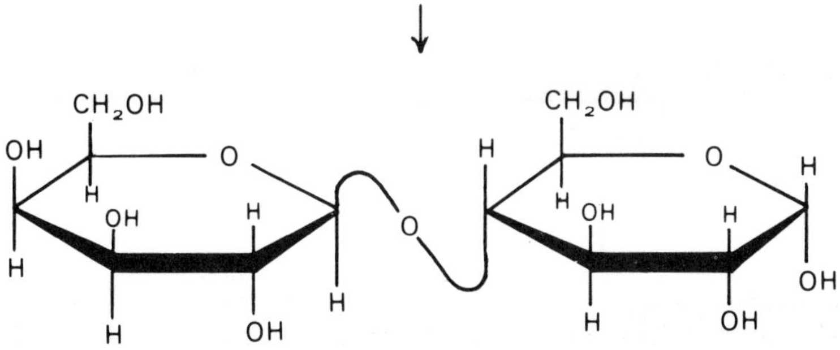

α-lactose

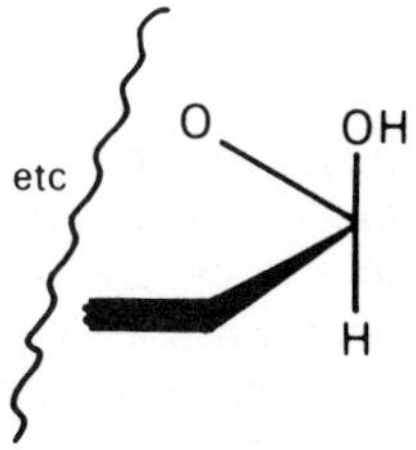

etc

β-lactose

In the case of lactose the glycosidic link is a β, 1–4
link as the galactose on the left is in the β-form.

Sucrose

Ordinary sugar is almost pure sucrose. Less refined sugar, found sometimes in overseas countries, is sweeter because it contains some invert sugar. Sucrose produced from sugar beet or from sugar cane, after refining, is exactly the same product. Most sugar in this country is now

Figure 1.14 Structure of sucrose

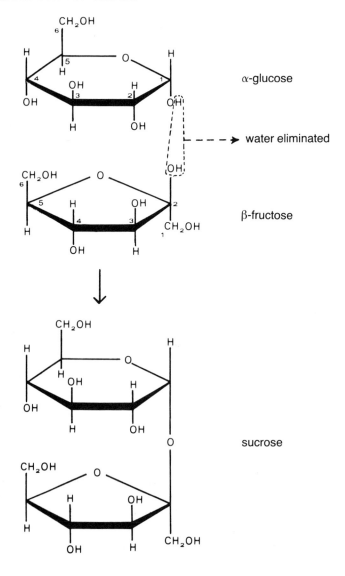

produced from beet which needs over twice the land area to produce the same amount of sugar as cane. The consumption of sugar, although less in recent years, is still at too high a level. Everyone is aware that too many sweets and too much sugar lead to tooth decay, but there may be more serious consequences of high sugar intake. It was thought that there was a direct correlation between sucrose intake and coronary heart disease, but this is now disputed. However, sucrose does cause an increase in blood fatty acids and high levels of these have been implicated in heart disease.

Sucrose is **not** a reducing sugar. Its structure is built up from α-glucose and β-fructose and there is only one sucrose because of the unusual linking of the two monosaccharides. In this combination the fructose has a five-sided structure. (The formula is difficult to remember, but it is easier to show the structure by placing the structure of glucose above that of fructose.)

The glycosidic link is an α, 1–2 link, which is unusual. Sucrose is the basis of a very large confectionery industry. Most types of confectionery are made from sugar which has been crystallized in a controlled manner from boiled syrup.

1.2.2 Non-sugars

1.2.2.1 Simple polysaccharides

Simple polysaccharides are long chains of one type of monosaccharide joined together. They are, therefore, big molecules and consequently insoluble in water. Usually these polysaccharides exist as long chains with their component monosaccharide units joined together with 1–4 glycosidic links (as we saw in the case of maltose). Occasionally there may be branches in the chains formed by 1–6 glycosidic links, and very occasionally by 1–2 or 1–3 links.

The general formula of the main simple polysaccharides in foods is $(C_6H_{10}O_5)_n$ where n can be many thousands of monosaccharide units. There are three polysaccharides of importance in this group; starch, cellulose and glycogen.

Starch

Starch is the energy reserve of plants. It is converted, when required, into sugars, particularly maltose and glucose. All plant storage organs are rich in starch, particularly seeds, tubers and unripe fruits. Starch is always accompanied by enzymes which can readily break it down and these can cause problems in food manufacture.

Starch exists in granules which are unique, in shape and size, to a particular plant source. It is possible by using a low power microscope to identify a particular type of starch by its shape and size. Some granules such as rice are small and angular whereas potato granules are large and more spherical.

Starch exists in two structural forms. The simplest form is amylose which is a straight chain of α-glucose units. Amylopectin (**not** to be confused with pectin) is composed of many shorter chains of α-glucose with many branches.

In many plants there is about four times as much amylopectin as amylose. However, the higher the amylose level the easier it is for a starch to gel, which is an important property of starch in cooking and food processing.

Table 1.2 Amylose levels in some starches

	Amylose content % (by weight)
Waxy mutant maize (corn)	0 – 6
Rice	16
Tapioca	18
Sweet potato	18 – 20
Potato	20 – 23
Wheat	22 – 25
'Steadfast' pea	65
Sugary mutant maize (corn)	60 – 70

Amylose is made up of about 300 glucose units joined together, but the number of units varies enormously. The α-glucose units are joined by α, 1–4 glycosidic links to make a chain, but very occasionally a branch may be formed particularly by an α, 1–6 link.

Figure 1.15 Structure of amylose

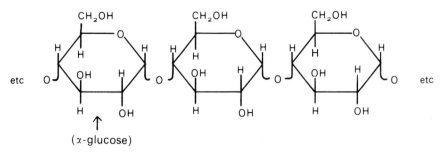

(α-glucose)

14

Remember that the glucose ring is in reality in the form of a 'chair'. This chair formation tends to cause the chain of glucose units to spiral and there are six glucose units per turn of the spiral.

Amylopectin is a more complex structure and is much larger than amylose, often having several thousand glucose unit components. The glucose units are linked to form short branching chains which again are in a spiral form. This tends to give amylopectin a tree-like appearance. The branches in the chains are produced by α, 1–6 glycosidic links.

Figure 1.16 Structure of amylopectin

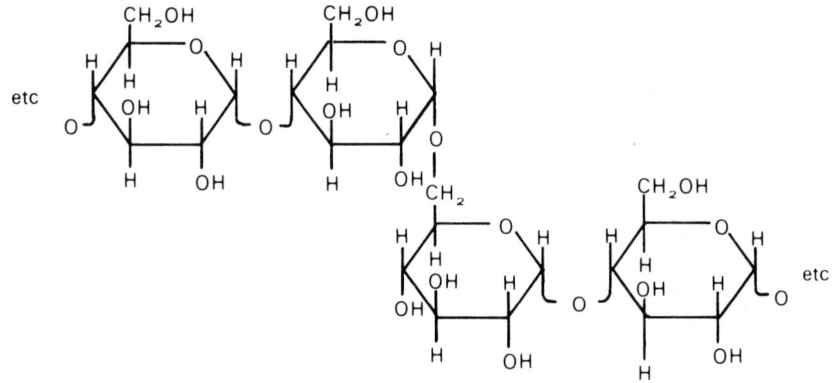

'tree-like' appearance:

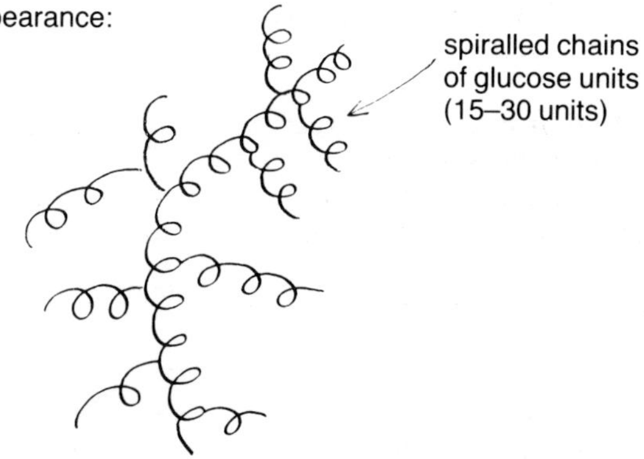

spiralled chains of glucose units (15–30 units)

Enzymic breakdown of starch

In plants starch is always accompanied by α- *and* β-*amylases,* known collectively as *diastase.* The two enzyme systems work together rapidly to break down starch to maltose and some dextrins. Although the enzymes work together it is easier to consider their actions separately.

α-*Amylase* is also known as liquefying or dextrinogenic amylase. This enzyme rapidly increases its activity during the germination of seeds. In malting of barley germination is encouraged for this reason. This amylase can split α, 1–4 links between the glucose units at random in the chain, thus breaking the chain down into small lengths known as dextrins (about 6–12 glucose units). The α, 1–6 links of amylopectin are by-passed as shown in Figure 1.17.

Figure 1.17 Action of α-amylase (simplified diagram)

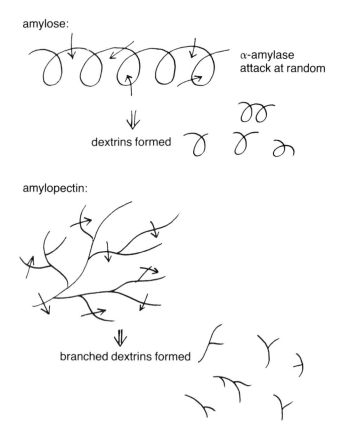

If a starch is used, for example, to thicken a soup when manufacturing canned soup it must be heat-treated to ensure no amylase activity is present. If α-amylase is present the starch-thickened soup will suddenly turn back to the viscosity of water as the enzyme breaks down the starch to dextrins.

β-*Amylase* forms maltose from both types of starch. However the enzyme can only attack starch in an organised manner by removing two glucose units, i.e. maltose, from the free ends of starch molecules. In the case of the straight chain amylose molecule the β-amylase will remove maltose units from both ends until all the amylose is converted into maltose. However, in the case of the branched amylopectin β-amylase is unable to convert completely the starch to maltose because the α, 1–6 links act as a barrier to the action of the enzyme. At each branch, therefore, β-amylase activity is stopped. In the case of amylopectin this will leave a central area in the tree-like molecule untouched by the enzyme. This part of the molecule remaining is called the 'β-limit' dextrin.

In reality as α- and β-amylase exist together the β-limit dextrin would be attacked by the α-amylase, thus releasing free-ends of the glucose chains for further attack by β-amylase. In this manner the starch will be almost completely broken down to maltose.

Figure 1.18 Action of β-amylase

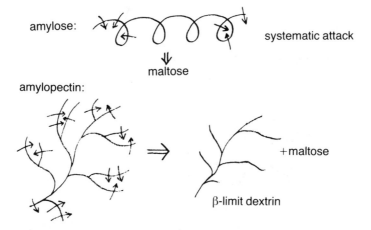

Gelatinization of starch

A very important property of starch is its ability to form a gel which is able to thicken a large number of foods. Gravies, soups, blancmanges, lemon pie fillings, custards and instant desserts rely on the gelatinization of starch.

When mixed with cold water starch granules do not dissolve, but just form a suspension. However, when the water is heated the viscosity of the mixture increases, and provided there is enough starch present, a gel is formed. Sometimes a firm gel does not form until the mixture is cooled again.

17

Starches gel at a certain temperature, called the *temperature of gelatinization,* but in reality the temperature is a range of temperatures and not usually fixed to a rigid temperature. Large starch granules, eg. potato, gel more easily and at a lower temperature than small densely packed granules such as those of rice. Amylose gels more easily than amylopectin, and starches rich in amylose consequently gel more easily. Gelatinization may be divided into three main stages:

Stage 1 – in cold water the starch takes up about 25% (of its own weight) of water.

Stage 2 – occurs at about 60° (depending on starch variety) and the granules swell rapidly, taking up between 3 and 10 times their weight in water.

Stage 3 – so much water has been absorbed (up to 20 times the weight of the granules) that the granules start to split. Starch molecules spill out into the surrounding water and its viscosity increases rapidly. The remaining starch granules stick together to form a three-dimensional network which on cooling forms a gel.

Importance of hydrogen bonding

In a starch in water suspension, as in stage 1, the starch granules are loosely bound together by hydrogen bonds. As the temperature rises molecular movement increases, thus rupturing the weak hydrogen bonds. Water molecules are able to penetrate gradually between the starch molecules. Thus, the starch granules swell, stage 2, until they burst open, stage 3. When the temperature falls again water molecules are trapped between the starch molecules. The water molecules effectively help bridge across from one starch chain to another, thus making the three-dimensional gel structure.

If a gel is formed from amylopectin it is with difficulty because of the tree-like structure of that type of starch. However, gels formed from mainly amylose, on standing, show a gradual contraction and the elimination of some of the bound water. This 'weeping' of water from the gel is known as *syneresis* and can sometimes be seen in canned products which have been thickened with starch or flour. The process overall is effectively the reverse of gelatinization, and is known as *retrogradation.* Starches which are mainly amylopectin are resistant to retrogradation. Starch manufacturers can modify starches rich in amylose by treating them to produce phosphate cross-bonded starches. These starches resemble amylopectin in having the ability to resist retrogradation but gel more easily like amylose.

It is possible for a manufacturer to dry a starch gel to produce what is known as 'pre-gelatinized starch'. On addition of cold water (or milk in some products) the dry powder gelatinizes instantly. This is the basis of some instant desserts and quick-cook products, such as some custard mixes.

Figure 1.19 Gelatinization of starch

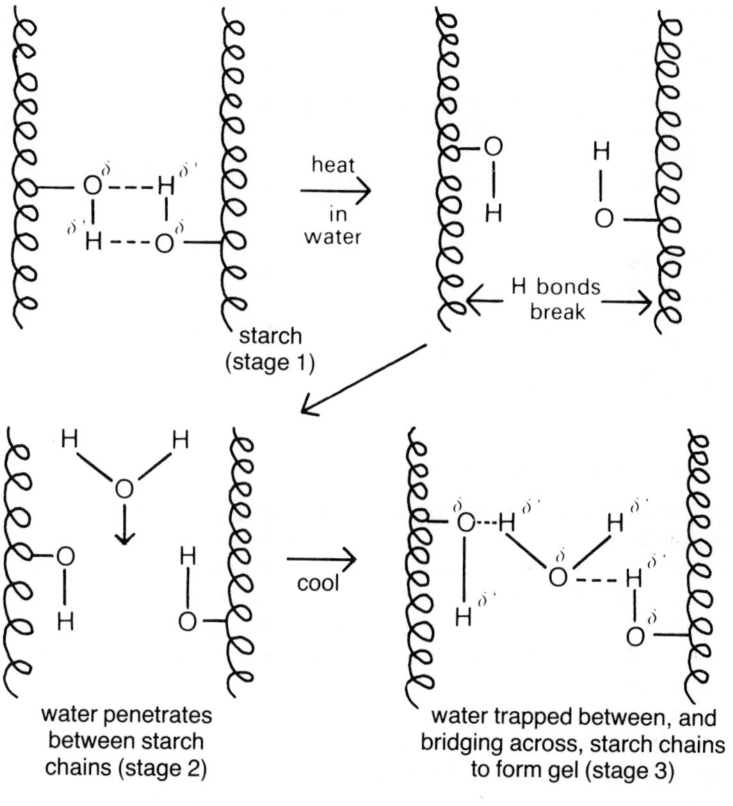

heat in water

starch (stage 1)

H bonds break

water penetrates between starch chains (stage 2)

cool

water trapped between, and bridging across, starch chains to form gel (stage 3)

Figure 1.20 Phosphate cross-bonded starch

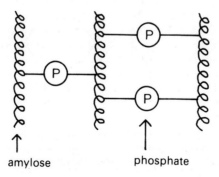

amylose

phosphate

Cellulose

Cellulose is an important constituent of the diet in supplying roughage or fibre. It has the ability to hold water and thus in the large intestine facilitates the muscular movements of the gut. Highly refined food is low in fibre, whereas 'primitive' foods tend to be rich in fibre. There is evidence, although much is still in debate, that high fibre protects against the 'diseases of civilization' such as obesity, diabetes, diverticular disease and cancer of the colon. There is also evidence that high fibre beneficially affects absorption of nutrients from the gut into the body and may help control blood cholesterol levels. Cellulose is only part of the fibre story, a number of other substances such as hemicelluloses, pectins and gums are involved.

Cellulose gives strength to plants in the form of long fibres. These fibres run in certain directions, particularly in stems, to resist wind for example. There is another type of cellulose which is amorphous and tends to absorb large amounts of water. This latter fact has been made use of in slimming foods. The cellulose swells in the stomach to give a 'full-feeling', but is not digested.

Cellulose is a very large molecule, often made up of several thousands of monosaccharide units. The monosaccharide unit in cellulose is β-glucose.

(*NB* It is important to notice the difference: starch is composed of α-glucose, cellulose is composed of β-glucose.)

The β-glucose units are joined together by β, 1–4-glycosidic links. In cooking and heat processing the rigidity shown by cellulose in plant tissues may be reduced, but cellulose is not generally affected, as for example starch is during its gelatinization.

Figure 1.21 Structure of cellulose

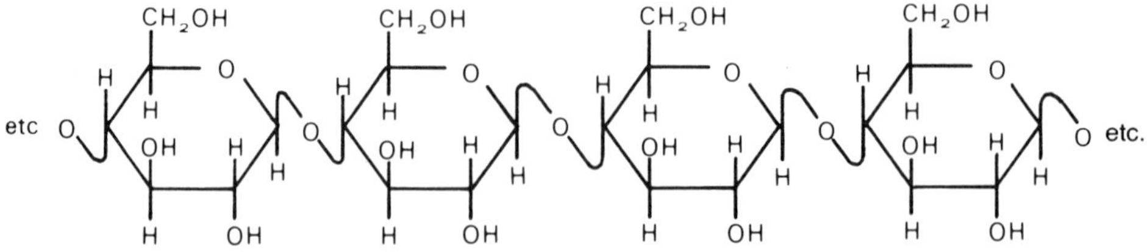

Hemicelluloses are complex mixture of substances which do not form fibres and are found in plants between their cells. They are often polymers of pentoses (5 carbon atoms) such as xylose.

Glycogen

Glycogen, like starch, is a reserve carbohydrate, but is found in animals. It is readily broken down to maltose and α-glucose, which is used up rapidly during muscular activity. Glycogen levels are important in muscle at the time of the animal's slaughter. Insufficient glycogen will produce meat of low acidity which leads to poor colour, poor texture and inferior keeping quality. (See Section 2.2.1.2.)

Glycogen is a large molecule closely resembling amylopectin, but with very short branched chains of about half the length of those found in amylopectin.

1.2.2.2 Complex polysaccharides

Complex polysaccharides are long chains, sometimes branched, of more than one type of monosaccharide joined together. Often these polysaccharides are built up of various derivatives of carbohydrates, particularly uronic acids.

Figure 1.22 Uronic acids and corresponding monosaccharide bases

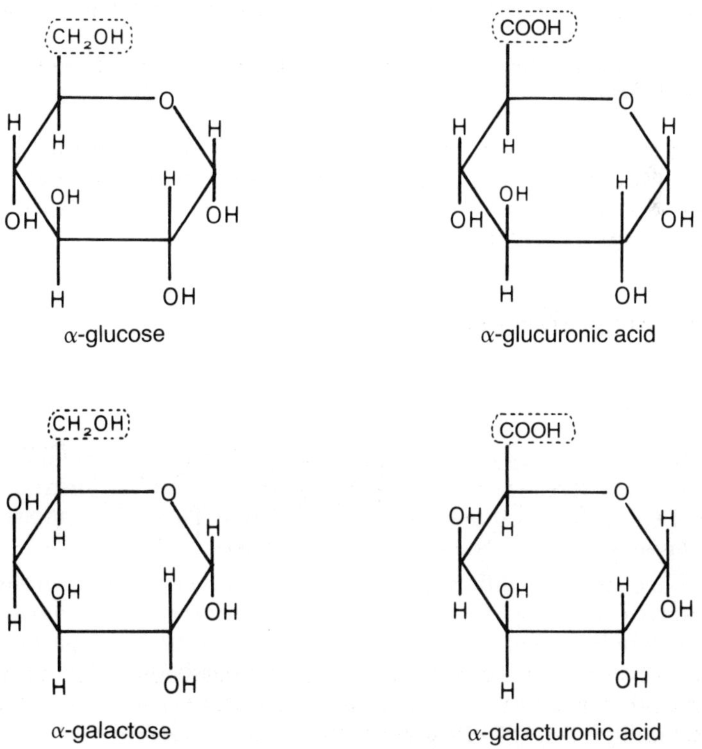

α-glucose

α-glucuronic acid

α-galactose

α-galacturonic acid

Pectin

Pectin is the most common example of a group of complex polysaccharides. Pectins are essential in jam making, but are also important in many other foods and for various functions. They occur naturally in the middle lamella between plant cells and also in cell walls. They are polysaccharides composed of α-galacturonic acid and its derivate methyl-galacturonate.

Figure 1.23 Structure of pectin

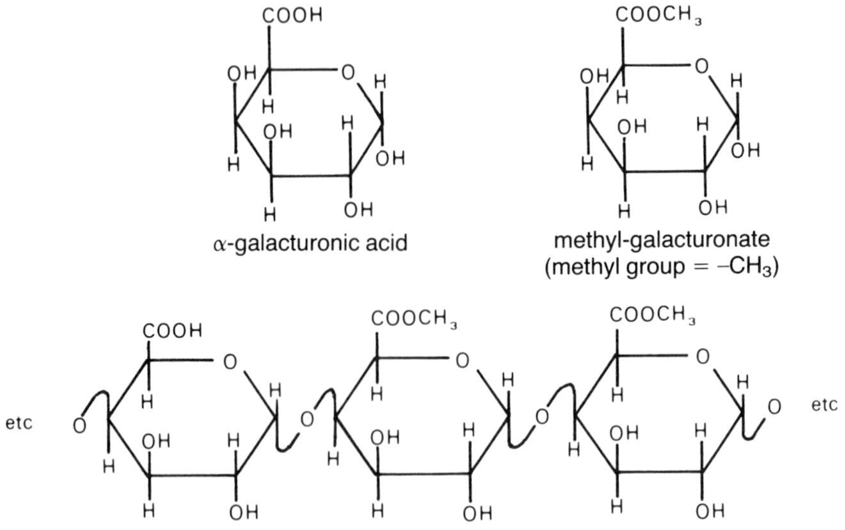

α-galacturonic acid

methyl-galacturonate
(methyl group = −CH₃)

The glycosidic link between the derivatives α-galacturonic acid and methyl-galacturonate is an α, 1–4 link.

(*NB* Note the link is different from that in starch as the right −OH group is at the top.)

Pectins are very variable in their composition. Chain lengths are variable and there is an infinite variation in the combination and order of each of the monosaccharide derivative units. Thus, it is impossible to obtain two pectins which are exactly the same.

There are several different types of pectins and their names are somewhat confusing. *'Pectin'* and *'Pectic substance'* are general names for the whole group. *'Protopectin'* is thought to be a parent pectic substance in plants from which other pectic substances are produced. *Pectinic acid* is pectin which contains a considerable number of the methyl-galacturonate units and is the pectin which forms a gel with sugar and acid in jams and other preserves. *Pectic acid* is made up of just α-galacturonic acids with few methyl galacturonates and it is incapable of gelling with sugar and acid. Unripe fruit is rich in pectin and on ripening the protopectin is broken down by enzymes to pectinic acids, then pectic acids, and the chains of units are broken. Thus a firm and crisp fruit becomes soft and juicy in its ripe form.

Pectic substances under the right conditions form gels, ie jams, jellies (not table jelly which is gelatine based) and preserves. Pectins derived from different sources vary widely in their jelly (gel) forming properties due to the different lengths of their chains and to the different numbers of methyl groups in their structures. A firm gel depends on:

(1) Amount of sugar
(2) Amount of pectin
(3) Molecular weight of pectin
(4) Number of methyl groups
(5) pH

A good gel is formed when there is sufficient sugar (about 65–68%). The sugar is thought to act as a dehydrating and orientating agent in bringing the pectin chains together and binding up water so that a three-dimensional gel structure can develop.

Figure 1.24 Gel structure of jam

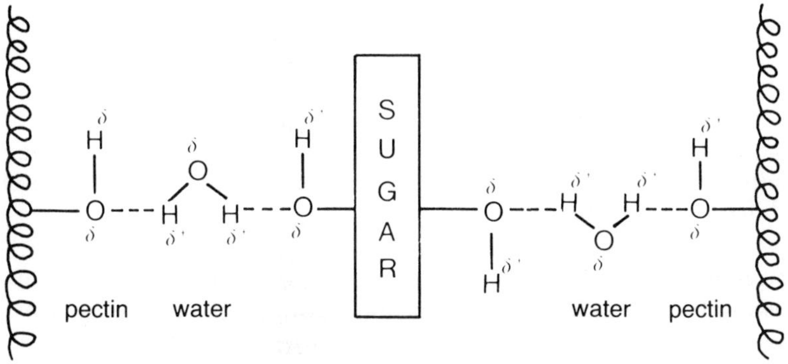

(– – –, H bonds)

Again, the importance of hydrogen bonding is seen in building up the gel structure. Some of the sugar is inverted by the natural acid of the fruit, usually under-ripe fruit, to invert sugar. Not only is invert sugar very sweet but it is very soluble in water. Any water will be dissolved in by invert sugar, thus making it unavailable as water-of-crystallization for the sucrose. Thus the high amount of sucrose does not crystallize out on standing but the jam stays as a smooth gel. Obviously, the pectin chains must be long enough to make a three-dimensional network possible. If there are not enough methyl-galacturonate units present then there must be too many α-galacturonic acid units which prevent the jam setting. This can occur when using ripe fruit. If the pH of the jam is not around pH 3·5, these α-galacturonic acids may ionize, thus:

$$\text{etc}-COOH \rightleftharpoons \text{etc}-COO^- + H^+$$

The charged carboxyl units $-COO^-$ will tend to force the pectin chains apart preventing gel formation. (Remember like charges repel each other.)

Boiling is carried out in jam making to bring the total dissolved solids, mainly sugar, to about 68%. Over-boiling will break down the pectin chains and effectively reduce their molecular weights. Low molecular weight pectins will not gel properly.

Some fruits, such as strawberries, are deficient in pectin, so a commercial pectin preparation is added. These are usually prepared from apple residues left from cider making. Some pectins, rich in methyl-galacturonate units, are rapid-set pectins. These pectins gel very rapidly and are useful for jams where it is desirable to suspend whole fruits or pieces of fruit in the jam.

Pectin is not always useful, for example, in some wines it may form a permanent haze. In some soft-drinks, such as bitter orange, pectin helps to support pieces in the drink and strengthens the *'permanent cloud stability'* of the drink.

Gums

Gums are a very diverse group of polysaccharides which have the ability to absorb large quantities of water and form firm gels under the right conditions. Used in foods in small amounts they often act as thickeners, stabilizers and emulsifiers.

Some plants produce gums on their stems or fruit when they are injured. Many of these gums are dried and sold commercially for use as thickeners or adhesives. Common examples of these are gum tragacanth, gum arabic, gum karaya and gum ghatti.

A number of important gums are produced from various seaweeds. Alginic acid, or its more common form sodium alginate, is produced from giant kelp. Another, similar gum is carageenan produced from Irish moss. Both of these gums are used widely as stabilizers in food products such as ice cream, syrups, processed cheese and salad dressing. Agar, produced from a red seaweed, has the ability to absorb enormous amounts of water and to form gels at very low concentrations. Although used as a stabilizer, and to replace gelatine in some confectionery, its main use is in the production of media for the cultivation of micro-organisms.

Review

1. Carbohydrates – $C_x(H_2O)_y$

2. Sugars – monosaccharides – glucose, fructose
   – disaccharides – maltose, sucrose, lactose

3. Non-sugars – simple polysaccharides – starch, cellulose, glycogen
   – complex polysaccharides – pectin, gums

4. A reducing sugar, eg glucose, breaks down Fehling's solution to give red precipitate

5. Sucrose – Non-reducing sugar

6. Optical activity – dextrorotatory, glucose   – hence dextrose (+)
   – laevorotatory, fructose   – hence laevulose (–)

7. Inversion of sucrose – sucrose →glucose+fructose

8. Invert sugar = glucose+fructose

9. Maltose = glucose+glucose, combined by α, 1–4 glycosidic link

10. Lactose = galactose+glucose, combined by β, 1–4 glycosidic link

11. Sucrose = glucose+fructose, combined by α, 1–2 glycosidic link

12. Starch – chains of α-glucose

13. Starch – 2 types – amylose – straight chains but in form of coil
    – amylopectin – highly branched 'like a tree'

14. Diastase = α-amylase+β-amylase
    Breakdown starch to dextrins and maltose

15. Starch gelatinization – faster with larger starch granules, easier with amylose

16. Hydrogen bonding of water important in gelatinization

17. Retrogradation opposite of gelatinization, water loss from gel known as syneresis

18. Cellulose – chains of β-glucose joined by β, 1–4 glycosidic links

19. Glycogen – chains of α-glucose similar to amylopectin

20. Pectin – complex polysaccharide – chains of galacturonic acid in various combinations with methyl-galacturonate

21. Pectin gels to form jelly, jams and preserves – depends on:
    amounts of sugar and pectin
    molecular weight of pectin
    number of methyl groups
    pH

Practical exercises: *Carbohydrates*

SAFETY – All experiments require care and some involve particular hazards. All chemicals, particularly acids and alkalis should be handled carefully, and ideally plastic gloves and safety glasses should be worn. The experiments are designed for simple laboratories or test kitchens with a minimum of specialized equipment.

1. Comparison of sweetness

Take sucrose as the standard at a value of 100, and compare the sweetness of a range of sugars, particularly glucose, fructose, lactose and maltose, giving a value to each.

2. Test for all carbohydrates – Molisch reagent

To a solution of a sugar in a test tube add a few drops of Molisch reagent (alpha naphthol in alcohol) and shake well. **Carefully** pour a little concentrated sulphuric acid down the inside of the tube and form a layer under the sugar solution. After standing a purple ring forms between the two layers for all carbohydrates.

3. Test for reducing sugars – Fehling's Test

Mix small but equal amounts of Fehling's 1 and 2 and add a small amount of sugar solution. Boil in a **water-bath** for 2 minutes (DO NOT BOIL IN A NAKED FLAME). A brick-red precipitate is formed, indicating a reducing sugar.

4. Test for a monosaccharide – Barfoed's Test

To a solution of a sugar (known reducing sugar) add excess of Barfoed's reagent. Mix and boil in a **water-bath** for 3 minutes. A monosaccharide produces a red precipitate and colouration.

5. Test for starch

Disperse some of the sample of starch into water and add a few drops of iodine solution. A dark blue/black colour indicates the presence of starch, but a lighter blue colour indicates a higher concentration of amylose and a brown/black colour the presence of higher amounts of amylopectin.

6. Microscopic examination of starch

Take a small amount of starch (maize, rice, potato, wheat, and as available) on the point of a knife, moisten with a drop of alcohol and then add a drop of water, stir and apply a drop to a microscope slide. Cover the drop with a cover-slip and press to ensure a thin layer of starch. A drop of iodine may be added to aid identification of the starch. Compare the different starch granules for shape, size and striations.

7. Gelatinization of starch

Make a mixture of starch in water (20% by weight). Heat the mixture and note the temperature. Continue to heat until the suspension thickens, record the temperature of gelatinization. Allow to cool and note any increase in viscosity of the starch gel.

8. Pectin gels

(**WARNING:** during boiling splashing may occur, which could result in painful burns.)

Special requirements: commercial preparations of pectin, buffer solutions from pH 2·5 to 4·0.

Preparation of pectin gel:

    Sugar 68 g
    Buffer solution 150 cm^3
    Pectin 0·8 g

Add the pectin to the buffer solution, stir well to dissolve. Gradually add sugar with stirring and heating. Boil for about 10 minutes (ideally to reach 68·5% total solids measured with a refractometer).

Variables:

1.   Sugar (total solids) – vary the sugar content, eg 50, 55, 60, 65 g.
     Note the effect on gel strength.

2.   pH – vary the pH by using different buffers from pH 2·5 to 4·0.
     Which pH gives the strongest gel?

3.   Pectin quantity – vary the amount of pectin, eg 2, 4, 6 and 8 g.

1.3 Lipids (fats)

Lipids form a very large group of vaguely connected compounds. The true fats are the largest sub-group of lipids and are the best source of energy in the diet. Lipids also include compounds which are vitamins, emulsifying agents, waxes, pigments and antioxidants. All lipids are soluble in organic solvents such as petroleum ether, chloroform or ethyl ether.

Lipids are very widely distributed in foods, even in foods not considered to be 'fatty'. Even fruits contain some lipid, in the form of 'cutin', for example, which gives the shine to apples. Beef, trimmed of fat, contains about 25–30% of lipid (on a dry weight basis). Nuts are a particularly good source of lipid, for example pecans, groundnuts, and walnuts can contain between 55 and 75% of lipid.

In addition to the naturally occurring lipid in a food, fat is often added during the cooking or preparation of the food. Not only do fats add flavour and richness but they are a good means of transferring heat in the cooking process.

There are a number of ways of classifying lipids, a simplified classification is given in Table 1.3. It is important to remember that oils and fats are chemically similar. However, oils are liquid at room temperature, whereas fats are solid. In hotter countries some fats will become oils and conversely some oils will be fats in colder countries.

Table 1.3 Classification of lipids

Type	Examples
Simple lipids	Natural fats Waxes
Complex lipids	Phospholipids
Lipoids	Steroids

1.3.1 Simple lipids

1.3.1.1 Natural fats

Fats are chemically placed in a class of substances known as *esters*. Esters are formed by the reaction of an alcohol and an organic acid. In natural fats the alcohol involved is always glycerol (glycerine). Glycerol is a more complex alcohol, compared with ethyl alcohol, in that it has three hydroxyl groups.

Figure 1.25 Structure of glycerol

(a)

$$\begin{array}{l} CH_2OH \\ | \\ CHOH \\ | \\ CH_2OH \end{array}$$

(b)

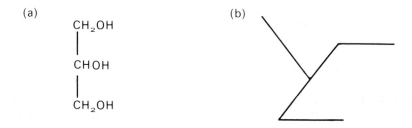

In reality glycerol takes the form of a 'bent tuning-fork' as shown in Figure 1.25(b).

The acids which react with glycerol can be represented as R–COOH, where –COOH is the acidic carboxyl group. The H of the –COOH will be removed, in effect, and combined with one –OH group from glycerol to form water. The rest of the acid will become attached to the glycerol structure as shown in Figure 1.26.

Figure 1.26 Formation of a simple triglyceride

$$\begin{array}{llll} CH_2\,OH & H\,OOC.R & & CH_2OOC.R \\ | & & & | \\ CH\;\,OH \quad + & H\,OOC.R & \rightarrow & CH.OOC.R \quad + \quad 3H_2O \\ | & & & | \\ CH_2\,OH & H\,OOC.R & & CH_2OOC.R \end{array}$$

The resulting ester formed in this reaction is called a *triglyceride*. If all three fatty acids are the same the triglyceride is called a *simple triglyceride*. Simple triglycerides usually do not occur in fats in foods, as the three acids reacting with glycerol normally are different. In this case a *mixed triglyceride* is formed, Figure 1.27.

Figure 1.27 Formation of a mixed triglyceride

$$\begin{array}{llll} CH_2\,OH & H\,OOC.R^1 & & CH_2OOC.R^1 \\ | & & & | \\ CH\;\,OH \quad + & H\,OOC.R^2 & \rightarrow & CHOOC.R^2 \quad + \quad 3H_2O \\ | & & & | \\ CH_2\,OH & H\,OOC.R^3 & & CH_2OOC.R^3 \end{array}$$

mixed triglyceride

These triglyceride molecules are randomly mixed in a fat but sometimes they occur in a more organised manner. In chocolate, for example, the triglycerides fit together to form stable fat 'crystals'. If the chocolate is not made satisfactorily then these triglycerides will rearrange themselves, particularly if the chocolate melts, and may separate out onto the surface to form a white 'bloom'.

All natural fats are *mixtures* of *mixed triglycerides*. Therefore, as glycerol is common to all triglycerides any differences between natural fats will be due to their different combinations of acids with glycerol. These relatively long chain acids are called *fatty acids*. Most fats contain a variety of fatty acids, but usually two or three dominate, and generally oleic acid is one of the main ones.

A list of common fatty acids is given in Table 1.4. Most of the common fatty acids have an even number of carbon atoms, but acids do occur rarely with odd numbers of carbon atoms. Some rarer acids may contain 30 carbon atoms.

Table 1.4 Common fatty acids

(a) Saturated fatty acids

No. of Carbon atoms	Name	Systematic name
4	Butyric	Butanoic
6	Caproic	Hexanoic
8	Caprillic	Octanoic
10	Capric	Decanoic
12	Lauric	Dodecanoic
14	Myristic	Tetradecanoic
16	Palmitic	Hexadecanoic
18	Stearic	Octadecanoic

(b) Unsaturated fatty acids

No. of Carbon atoms	Name	Systematic name	No. of double bonds
18	Oleic	9-octadecenoic	1
18	Linoleic	9, 12-octadecadienoic	2
18	Linolenic	9, 12, 15-octadecatrienoic	3

A saturated fatty acid has an acid group $-COOH$ and only carbon and hydrogen, with each carbon atom attached by single bonds to the next atom.

An unsaturated fatty acid contains at least one double bond between two adjacent carbon atoms. Remember that double bonds are a point of weakness, and not strength, in organic molecules as they can readily be broken by a wide range of substances.

Figure 1.28 Saturated and unsaturated fatty acids

Saturated fatty acid eg Stearic acid

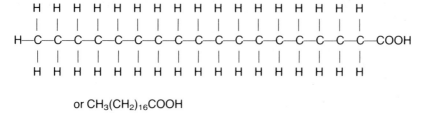

or $CH_3(CH_2)_{16}COOH$

or $C_{17}H_{35}COOH$

Unsaturated fatty acid eg Oleic acid

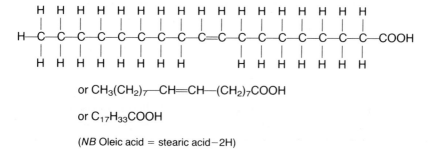

or $CH_3(CH_2)_7$—CH=CH—$(CH_2)_7COOH$

or $C_{17}H_{33}COOH$

(*NB* Oleic acid = stearic acid−2H)

The series of saturated fatty acids is summarised in Table 1.5. Note that the formulae increase by $-(CH_2)_2$ each time.

Fats containing only saturated fatty acids are hard and have melting points about room temperature. Oils contain unsaturated fatty acids and consequently have melting points below room temperature and thus are liquid. However, fats do not melt at a fixed temperature but over a range of temperatures. This property gives fats a unique *'plastic'* character. This plasticity results from the fact that fats are a mixture of mixed triglycerides of different melting points. As the temperature rises some triglycerides will melt whereas others remain solid. This allows the solid triglycerides to move within the fat and the fat is 'spreadable'.

Table 1.5 Formulae of saturated fatty acids

No. of Carbon atoms	Name	Formulae
4	Butyric	$CH_3(CH_2)_2COOH$ or C_3H_7COOH
6	Caproic	$CH_3(CH_2)_4COOH$ or $C_5H_{11}COOH$
8	Caprillic	$CH_3(CH_2)_6COOH$ or $C_7H_{15}COOH$
10	Capric	$CH_3(CH_2)_8COOH$ or $C_9H_{19}COOH$
12	Lauric	$CH_3(CH_2)_{10}COOH$ or $C_{11}H_{23}COOH$
14	Myristic	$CH_3(CH_2)_{12}COOH$ or $C_{13}H_{27}COOH$
16	Palmitic	$CH_3(CH_2)_{14}COOH$ or $C_{15}H_{31}COOH$
18	Stearic	$CH_3(CH_2)_{16}COOH$ or $C_{17}H_{35}COOH$

The *Iodine Value* is a measure of the degree of unsaturation, ie number of double bonds in a fat. A molecule of iodine will combine with each double bond in a fatty acid. Saturated fats have low iodine values, eg palm kernel about 15, whereas unsaturated fats have high values, eg groundnut oil about 90 (g of iodine per 100 g of fat).

Hydrogenation

Some oils are so unsaturated that they are of little use in their natural state. For many years the process of *hydrogenation* or hardening has been undertaken to remove some of the double bonds in the fatty acids and effectively to make them more saturated. Hydrogenation changes a liquid oil into a solid fat by adding hydrogen across the double bonds in the unsaturated fatty acid molecules. The oil is heated and stirred with a small amount of nickel which acts as a catalyst. The nickel is a surface active catalyst and has 'active sites' where hydrogen atoms are taken up by the unsaturated fatty acids, as shown in Figure 1.29.

Figure 1.29 Hydrogenation of an unsaturated fatty acid

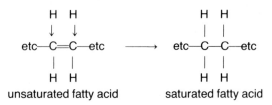

Oils and fats transfer heat well to foods being cooked, but their usefulness is limited as heat will cause their breakdown. Triglycerides will decompose on heating by a number of methods. Glycerol may be split from the component fatty acids and then converted to acrolein which may appear as an unpleasant smelling blue smoke. Unsaturated fatty acids are susceptible to oxidation which leads to rancidity (see Section 1.3.4). The old type of copper pans and utensils greatly accelerated this problem. Recent work has shown that the repeated heating of some vegetable oils may cause the accumulation of carcinogenic substances in the oil. In general, most fats and oils are suitable for frying. In the case of butter and margarine however, the water present (up to 16%) restricts the rise in temperature during frying. Rapid discolouration and breakdown of the triglycerides occurs, usually accompanied by spluttering.

1.3.1.2 Vegetable oils

Plants are the major source of oils and fats used in food processing and cooking. Something like 70% of all fat originates in plants. The oils are produced from the seeds of several hundred different varieties of plant. The principle one is soya bean, accounting for nearly 20% of all fat production. Other sources are groundnut (peanut or monkey nut), rapeseed, cottonseed, sunflower, sesame, coconut, palm and palm kernel. These oils are used in margarine, cooking fats and oils, ice cream and salad dressing. Vegetable oils are usually resistant to rancidity as they contain natural antioxidants.

Soya bean

Although the soya bean has been widely cultivated in the Far East the major producer is the USA, but new varieties are being developed which are capable of growing in colder countries. The soya bean prefers a hot and damp climate. The beans yield between 13 and 20% oil. Soya bean oil is a major component of many margarines and cooking fats.

Groundnut

The groundnut has a number of other names which include peanut, earth-nut and arachis nut. Groundnut oil is often called arachis oil. It is produced in many tropical countries and in some sub-tropical areas. The oil content of groundnuts is often as high as 45%. Like soya bean oil, groundnut oil is used in the manufacture of cooking fats, margarine, ice cream and salad dressings.

Sunflower

The production of sunflower oil is increasing worldwide, particularly in sub-tropical areas. Normally the seed contains between 20 and 30% of oil, but new varieties are yielding up to 40%. It is used in a similar manner to soya and groundnut oil and also, in a less refined form, in soaps and paints.

Rapeseed

Rapeseed or Canola, as it is now called in Canada, is the European answer to the American soya bean oil. An alternate source of vegetable oil, which could be grown in Europe, was sought to offset the very large imports of soya bean oil. The heavily subsidised production of rapeseed shows itself in the attractive blocks of yellow fields found in Eastern England and in parts of Europe. Rapeseed yields between 35 and 40% oil. However, early varieties contained a fatty acid known as erucic acid (22 carbon atoms and one double bond) which has been shown to cause heart disease. Varieties have been developed subsequently which are low or free of this fatty acid.

Oil extraction

Two methods of extraction are employed in removing oil from plant sources. The oil is expressed from the seeds by physical pressure in some form of press. Instead of this method, or often as well, the oil is extracted by the use of organic solvents. The seed is first cleaned, stones and any metal are removed. The seed is then broken between heavy rollers and is reduced to a coarse meal. The meal of broken seeds is heated (70–110°C) and stirred in a large vessel. The cells of the seeds burst open and release their oil. The oil is then extracted or expelled from the seed in an expeller or screw press (similar to a mincer). Considerable pressure is exerted, but about 5% of the oil remains in the seed cake and has to be removed by solvent extraction. Petroleum ether is used to wash the oil seed, which is firstly flattened into flakes. Modern continuous solvent extraction systems employ a counter-current method whereby seed moves in one direction and the solvent in the opposite direction. Fresh solvent is used at the end of the process to extract the last traces of oil and solvent which has already dissolved some oil is used initially to remove the easily dissolved oil from the seed flakes. The solvent is then distilled off to leave a 'crude' oil which needs refining.

Refining

The oil usually has many impurities such as free fatty acids (odour), pigments, waxes and other material. Refining usually has three stages. Firstly, any free fatty acids must be neutralized. Caustic soda is used which reacts with fatty acids to make a soap. This soap settles to the bottom of the oil which is then washed with warm water several times. The oil is bleached to remove pigments and other colours. Bleaching is carried out by using Fuller's earth or charcoal, which is added to hot oil under vacuum and stirred for about 15 minutes. The Fuller's earth is filtered out to leave a clear oil.

The third process is deodorization which is carried out by passing steam through the hot oil under high vacuum. Odoriferous substances become volatile and are removed leaving a clear, odour free oil. The oil may be used in this form, but often it is hydrogenated and blended with other fats to make margarine and cooking fats.

1.3.1.3 Animal fats

Fat in animals is found mainly in the adipose tissue around organs such as the kidneys and heart, and under the skin. Fat is extracted by the traditional process of rendering which involves gently heating the animal tissue to melt the fat which then easily separates from the protein material. The two principal fats from animals are lard and butterfat. (Please remember that butter itself is not a pure fat, but an emulsion of water in fat.) Both butterfat and butter will be discussed in the section on dairy products (see Section 2.1).

Lard is prepared by rendering fat from pigs. It has declined in popularity because of the growth of hydrogenated vegetable oils and specially prepared fats. It has been used extensively in the past as a shortening agent, but will not cream well. For cake-making good creaming ability is essential. The process of *interesterification* was developed to improve the

Figure 1.30 Interestification

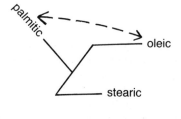

(a possible triglyceride in lard)

heat with
sodium ethoxide (a catalyst)

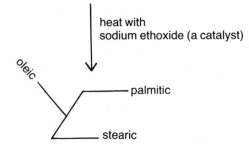

creaming ability of lard. The process improves the plasticity of lard and reduces any graininess. Natural lard tends to be made up of large 'crystals' of triglycerides with two saturated fatty acids and one unsaturated fatty acid. These saturated acids are generally stearic and palmitic acids and one is always found in the middle position of the triglyceride molecule. This is the cause of the poor creaming ability of lard. Interestification causes fatty acids to change their positions and the number of triglycerides with these acids in the centre position is greatly reduced. The lard is heated to about 105°C in the presence of a catalyst (sodium ethoxide) to achieve this change.

The consumption of animal fats has in the last few years declined considerably, and has been matched by an increase in the consumption of vegetable oils and fats.

1.3.1.4 Fish oils

Pelagic fish, eg herring, pilchards, sardines and mackerel contain up to about 20% of fat. Spawning reduces this figure dramatically. The fat is not stored in adipose tissue but is distributed throughout the tissues. Fish oils contain a very high percentage of unsaturated fatty acids. Some of these acids may have up to six double bonds. This is reflected in the high iodine value of marine oils, eg herring oil has a value of 190–200. The double bonds are readily oxidised which leads to rapid rancidity in this type of oil. Before fish oils can be used they must be processed and hydrogenated. They are usually rich in a number of vitamins particularly A and D. For many years fish oils have been used as vitamin supplements, for example cod-liver oil.

1.3.1.5 Monoglycerides and diglycerides

In Section 1.3.1.1 the structure of a triglyceride was discussed and involved the esterification of glycerol with three fatty acids. In a similar manner it is possible for glycerol to combine with only one fatty acid. In this case a monoglyceride is formed. A common example of such a monoglyceride is glyceryl monostearate (GMS) which is formed from glycerol and one molecule of stearic acid.

Figure 1.31 Formation of glyceryl monostearate

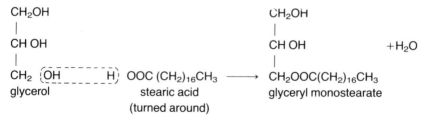

Looking at the structure of GMS it is obvious that the hydroxyl groups (–OH) are not esterified with a fatty acid. Because of this fact, GMS has the useful property of acting as an emulsifying agent. If an oil is mixed with water and stirred it forms a temporary emulsion but then rapidly separates onto the surface of the water. If a very small amount of GMS, or any other emulsifying agent, is added the separation of oil is prevented or considerably retarded. GMS has the ability to surround oil droplets and prevent them coalescing to form larger droplets which rise to the surface ever more rapidly. The hydroxyl groups of the GMS molecule are soluble in water, whereas the stearate part of the molecule is soluble in oil. However, the hydroxyl groups will not dissolve in, and in fact will repel, any fat droplets. In this manner the fat droplets are completely surrounded by GMS and cannot combine, and therefore, cannot rise to the surface of the water. The process of emulsification is represented diagrammatically in Figure 1.32. Other emulsifying agents, or emulsifiers, behave in a similar manner and so do soaps and detergents.

Figure 1.32 Emulsification of oil in water

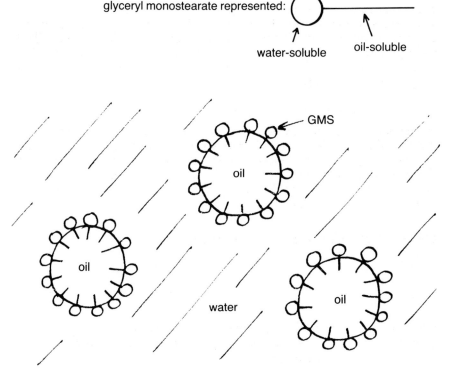

Diglycerides are formed in a similar manner to monoglycerides, but obviously two fatty acids combine with glycerol and only one free hydroxyl group remains. Diglycerides also act as emulsifiers but tend to dissolve more in the fat than water because of the two fatty acids combined with the glycerol. Blends of emulsifiers are used in the food industry in a large range of products where it is necessary to prevent fat separation, for example in margarine, ice cream, syrups, salad dressing and desserts.

1.3.1.6 Waxes

Although classified under simple lipids the waxes are complex mixtures of a range of substances. They are usually alcohols which have long chains of 18 to 22 carbon atoms joined to a range of fatty acids such as palmitic, stearic, oleic and linoleic. Waxes line the intercellular air spaces in plant leaves and stems and also are found on the surface of fruits, such as apples. This surface wax or 'cutin' limits water loss by transpiration from the fruit and also prevents water entering the fruit. An apple placed under a running tap clearly demonstrates this.

1.3.2 Complex lipids

A large group of naturally occurring substances are formed from glycerol, and other alcohols or sugars, combined with phosphoric acid and a range of alkaline or basic substances. The most important group is the *phospholipids* of which the *lecithins* are very significant in foods. The general structure of a lecithin is given in Figure 1.33. Like the monoglycerides and diglycerides, lecithins act as emulsifiers because they have fat-soluble and water-soluble parts to their structures.

Figure 1.33 Lecithin

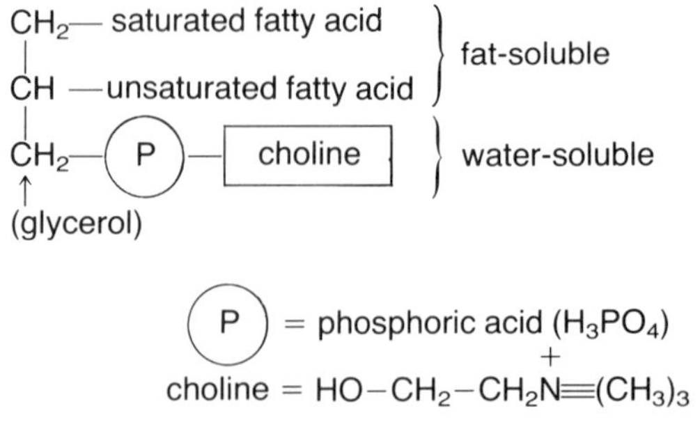

$$CH_2 - \text{saturated fatty acid}$$
$$CH - \text{unsaturated fatty acid} \quad \} \text{ fat-soluble}$$
$$CH_2 - (P) - \boxed{\text{choline}} \quad \} \text{ water-soluble}$$
(glycerol)

(P) = phosphoric acid (H_3PO_4)
+
choline = $HO - CH_2 - CH_2N \equiv (CH_3)_3$

Lecithins are thus important natural emulsifying agents found in many foods such as milk, egg yolk and vegetable oils. The best commercial source is from the soya bean.

1.3.3 Lipoids

The group of substances known as *steroids* are classified as lipids, or lipid-like, hence *lipoid*. The lipoids could also include the carotenoids and tocopherols, but these are dealt with in Sections 1.6.2.1, 1.6.2.3 and 1.7.1.1. The steriods are not fatty acid derivatives and include *cholesterol*, vitamin D and some bile acids.

Cholesterol occurs widely in animal fats, and the human body produces up to twice its own requirement for use in manufacture of vitamin D_2 and a number of hormones. However, up to about 25% of blood cholesterol comes from the diet. Butter contains about 250mg/100g whereas egg yolk solids are as high as 3·5% cholesterol. During the refining of oils steroids are often removed during the treatment with caustic soda. However, many vegetable oils contain some steroid generally known as *phytosterol*. The structure of cholesterol is given in Figure 1.34 and it can clearly be seen to be totally different from other lipid structures.

Figure 1.34 Cholesterol

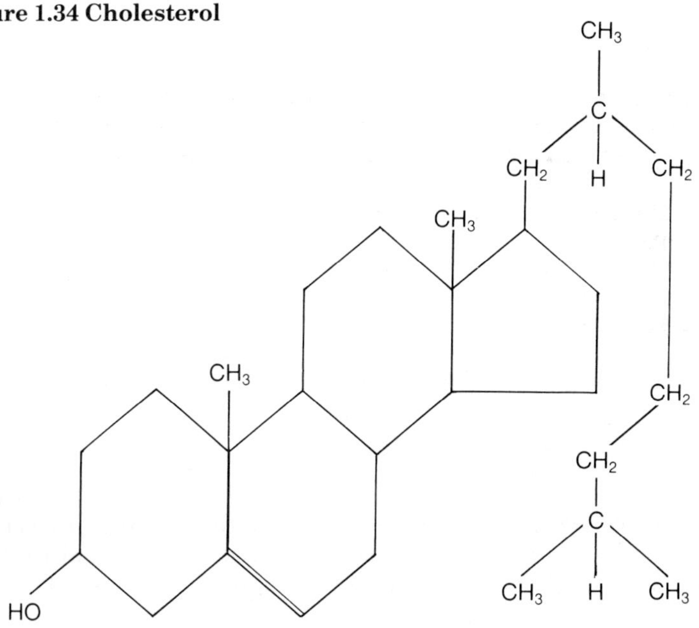

It is conventional to miss out just H
on formulae such as this,
but all C atoms must have a valency of 4.

1.3.4 Rancidity

The flavour of rancid fat is well known to everyone, but some people, particularly from hotter countries, have grown to prefer some degree of rancidity in their fatty foods. Two types of rancidity occur in fats and these are termed *hydrolytic* and *oxidative rancidity*.

Hydrolytic rancidity is the less common of the two overall but is quite common in emulsion systems such as butter, margarine and cream; it also occurs in nuts and some biscuits. Water must be present for hydrolytic rancidity to occur as, in effect, it is the reverse process of the esterification of glycerol with three fatty acids. The triglycerides are hydrolysed and the three fatty acids are set free. In the presence of water alone, the process is very slow but certain enzymes generally called lipases or lipolytic enzymes (split fat) greatly accelerate the process. Some bacteria produce these enzymes and so will cause this type of rancidity if they contaminate a certain fat or food. The free fatty acids produced by this hydrolysis often have unpleasant flavours. This is particularly so with the short chain fatty acids, but flavour disappears above about 14 carbon atoms. Hydrolytic rancidity in butter yields the dreadfully rancid smelling butyric (butanoic) acid. However this rancidity is actually desirable in the production of certain cheeses, particularly blue cheeses, as without the release of free fatty acids the cheese does not develop its full flavour. Fats should be heat treated to kill any micro-organisms and inactivate any enzymes present.

In foods, *oxidative rancidity* or *autoxidation* is by far the most important type of fat deterioration. Fats, and oils particularly, slowly take up oxygen over a period of time and then the flavour of rancidity is eventually detected. At this stage rancidity is well advanced and cannot be reversed. This uptake of oxygen, and the chain of events which it starts, is related to the unsaturation of the fat, ie the number of double bonds in the constituent fatty acids. Hard fats are resistant to this oxidation as they contain fewer double bonds, whereas fish oils being highly unsaturated are very susceptible. The oxidation of fats takes place by means of a chain reaction involving the production of highly reactive particles called *free radicals*. Normally the bonds between carbon and hydrogen atoms are produced by the sharing of electrons (covalent bonds). Some outside energy source, for example, ultra-violet light, splits these shared electrons into single electrons as shown in Figure 1.35.

The production of free radicals is initiated by various energy sources such as heat, light, particularly UV light, traces of metals, eg copper and iron, and some peroxides. Obviously, to prevent fat becoming rancid these initiators must be kept away from stored fat. Oils should be stored in glass lined containers as iron vessels initiate oxidative rancidity.

Figure 1.35 Free radical production—initiation phase

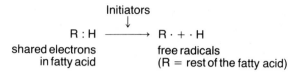

Initiators
↓

R : H ⟶ R · + · H

shared electrons　　free radicals
in fatty acid　　(R = rest of the fatty acid)

This production of free radicals in the fat is called the *initiation phase*. Oxygen is now taken up by the fat as the free radicals combine with the oxygen to make peroxide radicals (ROO·). This second phase is called the *propagation* phase. Peroxide radicals attack other fatty acids and produce yet more free radicals, so the process accelerates.

Figure 1.36 Propagation phase

R · + O_2 ⟶ ROO ·

free radical　　peroxide radical

then:　RH + ROO · ⟶ ROOH + R ·

another fatty　　hydro-　a new
acid　　peroxides　free radical

The hydroperoxides (ROOH) produced are unstable and break down to produce alcohols, aldehydes and ketones. These compounds provide the odour of rancid fat. The process continues at an ever increasing rate until all the fat becomes rancid, or no further oxygen is available, or the free radicals start to react with each other. The last phase is the *termination phase*.

Figure 1.37 Termination phase

(interaction of free radicals to make stable products)

R · + R · ⟶ R : R

ROO · + R · ⟶ ROO : R

Oxidative rancidity is an irreversible process of deterioration which can only be retarded and not prevented completely.

Obviously metals and light should be kept away from fat, which must be kept cool and free from oxygen or air. *Antioxidants* can be used which have the ability to absorb oxygen or to prevent the formation of the free radicals by forming stable radicals. The use of the additives is controlled by regulations. Table 1.6 lists antioxidants currently permitted in the European Community. Some foods, particularly vegetable oils and foods containing some spices, eg cloves, show marked resistance to rancidity. Salt tends to accelerate rancidity, whereas sugar, in biscuits for example, retards the deterioration.

Table 1.6 Permitted antioxidants

Antioxidant	E number	Applications
Form stable free radicals:		
Butylated hydroxyanisole (BHA)	E320	biscuits, stock cakes
Butylated hydroxytoluene (BHT)	E321	chewing gum
Tocopherol extract (Vit. E)	E306	vegetable oils
Synthetic alpha-tocopherol	E307	cereal-based baby foods
Synthetic gamma-tocopherol	E308	,, ,, ,,
Synthetic delta-tocopherol	E309	,, ,, ,,
Absorb Oxygen:		
L-ascorbic acid	E300	fruit drinks, dried potatoes
Ascorbyl palmitate	E304	scotch eggs
Sodium L-ascorbate	E301	meat loaf, sausages
Calcium L-ascorbate	E302	scotch eggs
Propyl gallate	E310	margarine, vegetable oils
Octyl gallate	E311	,, ,, ,,
Dodecyl gallate	E312	,, ,, ,,
Lecithins	E322	low fat spreads (also emulsifier)
Ethoxyquin	No E No.	prevention of 'scald' discolouration in apples and pears

1.3.5 The fat debate

The Western diet is too rich in fats and it is generally agreed that this level of consumption should be significantly reduced. Animal fats, as they are rich in saturated fatty acids and cholesterol, have been heavily cristicised over a number of years resulting in declining sales.

Cholesterol and fatty acids form deposits or plaques on arterial walls and obviously this reduces the size of the artery, making it harder for the heart to pump the blood through the blood vessels. At a minimum this leads to higher blood pressure, but often to heart disease and death. However, the connection between high cholesterol foods, such as eggs, and this deposition on the arterial walls is not conclusive. Although it is wise to avoid such foods the deposition may well still occur from the cholesterol synthesised in the body. Diets rich in fats have recently been implicated as causative agents of cancer, particularly breast and colon cancers.

In the chapter on carbohydrates it was pointed out that sucrose increases blood fatty acids, whereas high fibre diets decrease them. Modern life styles, involving stress, and the lack of exercise may also increase the problem. However, recent work has shown that a slight to moderate intake of alcohol is beneficial in removing some arterial cholesterol and fatty acids.

Some people are fortunate in having good fat translocation systems in the blood, others have inherited poor systems and are likely to develop problems of this nature.

The consumption of polyunsaturated fatty acids, eg in soft margarine, has also been shown to lower blood cholesterol levels. Until recently little attention has been paid to what are termed 'essential fatty acids'. The most important of these fatty acids is linoleic acid, also linolenic is essential, but the third essential acid, arachidonic, can be synthesised from the other two in the human body. These essential fatty acids are involved early in the human embryo in the construction of brain and nervous tissue. Deficiency of the acids has been implicated in skin disorders and in the development of multiple sclerosis. However, more recently investigations have shown that they help to reduce heart disease.

Evidence has mounted that the diet of many people in Britain is deficient in essential fatty acids. Fatty fish, such as mackerel, are the only rich source. The Japanese are less vulnerable to heart attacks than Westerners and their diet, rich in fish, rice and vegetables, seems to be the reason. Eskimos, similarly, have a diet rich in fish oils, and are reputed to suffer very little heart disease.

So, current knowledge suggests that we should consume less fat, but more fish oils at the expense of animal fats, have plenty of exercise and eat plenty of roughage (dietary fibre).

Review

1. Lipids: Simple lipids, eg natural fats, waxes
   Complex lipids, eg phospholipids
   Lipoids, eg steroids

2. Natural fats – mixtures of mixed triglycerides

3. Mixed triglycerides → glycerol + 3 different fatty acids

4. Glycerol – structure similar to 'bent tuning-fork'

5. Saturated fatty acids – acid group ($-COOH$) plus carbon and hydrogen sometimes in long chains

6. Unsaturated fatty acids – contain at least one double bond between two adjacent carbon atoms

7. Common saturated fatty acids include stearic, palmitic and myristic

8. Most common unsaturated fatty acid is oleic

9. Essential fatty acids are linoleic, linolenic and arachidonic

10. Iodine value – measure of unsaturation, ie number of double bonds in a fat

11. Hydrogenation – eliminates double bonds by adding hydrogen, this hardens a fat or oil

12. Interesterification — technique to rearrange fatty acids attached to glycerol

13. Monoglyceride = glycerol+one fatty acid
    Diglyceride   = glycerol+two fatty acids
                  – act as emulsifying agents, eg glyceryl monostearate

14. Waxes – esters of long chain alcohols and fatty acids

15. Phospholipids – eg lecithin = glycerol+two fatty acids
                                  +phosphoric acid+base eg choline
                  – act as naturally occurring emulsifying agent

16. Lipoids – eg steroids, such as cholesterol

17. Body produces about twice its requirement of cholesterol

18. Accumulation of cholesterol in blood vessels not necessarily connected with diet

19. Rancidity – hydrolytic – caused by enzymes
              – oxidative – caused by presence of oxygen, metals eg copper,
                UV light and heat
              Requires unsaturated fatty acids

20. Antioxidants – absorb oxygen or stop complex reactions in oxidative rancidity

Practical exercises: *Lipids*

SAFETY – all organic solvents should be treated as highly inflammable and correspondingly should be used away from naked lights and flames.

1. Grease-spot test

Take the food sample and wrap a small amount in a filter paper. Allow to dry; a translucent ring on the paper around the food material indicates the presence of lipid.

2. Sudan staining of lipids

The dye Sudan III or IV will stain traces of lipid a red colour. The food material must be broken up in some way; the dye is added to the food. Lipid will absorb the dye and take up the red colour.

3. Solubility of lipids

(**SAFETY:** work in fume cupboard or in ventilated area)

Use a range of organic solvents as available, and also water. Place a small amount of fat in a test tube and add a solvent. Repeat with different solvents with a fresh sample of fat. Observe the solubility of lipid in each solvent.

4. Use of emulsifiers

Stir a sample of oil in water and note the speed of complete separation of the oil. Add a small amount of detergent or soap and note the emulsification of the oil in the water. Repeat using commercial emulsifiers, such as glyceryl monostearate, and lecithin, and also repeat using egg yolk.

5. Acrolein formation

Acrolein is produced by dehydration of glycerol and gives the smell to burnt fat. Heat a few drops of glycerol (glycerine) with a dehydrating agent, eg anhydrous calcium chloride; note the smell of acrolein. Repeat with a number of oils and fats.

6. Melting point of fats

Take a sample of fat and melt it, then dip the end of a capillary tube into the molten fat. Cool in a freezer to solidify the fat; then attach the tube to the bottom of a thermometer. Gently heat the thermometer and tube in a warm bath. Note the temperature at which the fat becomes clear and moves in the tube.

1.4 Proteins

The molecules of protein are the largest known and are responsible for growth, repair and maintenance of the body. The richer countries of the West, and particularly of the Northern Hemisphere, consume considerably more protein than poorer Third World countries. Protein deficiency diseases are widespread in poor countries. In addition to containing the elements carbon, hydrogen and oxygen, protein also contains nitrogen and occasionally sulphur or phosphorus.

Proteins are large polymers built up of units of *amino acids*. Although over eighty amino acids exist only about twenty are found in food protein. Amino acids always contain an amino group ($-NH_2$) and a carboxyl group ($-COOH$). Some amino acids may contain more than one of these groups and as we shall see this will affect their properties.

The general formula of an amino acid is given in Figure 1.38.

Figure 1.38 General formula of an amino acid

$$NH_2-\overset{\overset{\displaystyle H}{|}}{\underset{\underset{\displaystyle R}{|}}{C}}-COOH$$

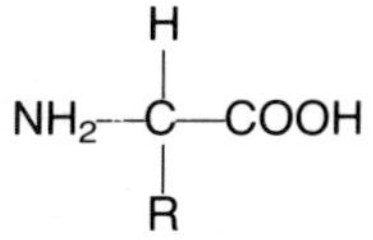

(The R group will vary in each amino acid, see Table 1.7)

45

The simplest amino acid is when R = H, this acid being amino acetic acid or glycine. In Table 1.7 the twenty amino acids are listed with their formulae. If the amino acid contains one amino group (alkaline) and one carboxyl group (acid) then it is neutral. More than one amino group gives

Table 1.7 Common amino acids

	R group =
Neutral:	
Glycine	$H-$
Alanine	CH_3-
Valine	$(CH_3)_2CH-$
Leucine	$(CH_3)_2CH\,CH_2-$
Isoleucine	$CH_3CH_2\,CH(CH_3)-$
Norleucine	$CH_3(CH_2)_3-$
Phenylalanine	$C_6H_5CH_2-$
Tyrosine	$C_6H_5(OH)CH_2-$
Serine	$HO\,CH_2-$
Threonine	$CH_3CH\,(OH)-$
Cysteine	$HS\,CH_2-$
Cystine	$HOOC\,CH\,(NH_2)\,CH_2\,S_2\,CH_2-$
Methionine	$CH_3\,S\,CH_2\,CH_2-$
Tryptophan	
Basic $(2-NH_2\,\text{groups})$	
Ornithine	$H_2N\,(CH_2)_3-$
Arginine	$\begin{array}{c} NH_2 \\ | \\ HN = C-NH\,(CH_2)_3- \end{array}$
Lysine	$H_2N\,(CH_2)_4-$
Histidine	
Acidic $(2-COOH\,\text{groups})$	
Aspartic acid	$HOOC\,CH_2-$
Glutamic acid	$HOOC\,(CH_2)_2-$

the amino acid an alkaline or basic character; similarly, more than one carboxyl group makes the amino acid more acidic. The amino acids have been sub-divided according to these features in the table.

Amino acids can be further sub-divided into essential amino acids (indispensable in the diet) or non-essential amino acids (which the body can synthesise sufficiently). There are eight essential amino acids. These are:

Valine	Threonine
Leucine	Methionine
Isoleucine	Tryptophan
Phenylalanine	Lysine

Children also require histidine during rapid growth. Arginine may also be considered as essential as it is only synthesised slowly in the body.

Proteins in foods may be classified on their amino acid content. Proteins that contain all the essential amino acids, for example meat proteins, in proportions capable of promoting growth are described as *complete proteins* or proteins of *high biological value*. These proteins used to be called *first class proteins* and proteins deficient in one or more amino acids were *second class* eg plant proteins. However, as any vegetarian will quickly point out, it is very simple to mix foods of plant origin to ensure that the meal contains all the essential amino acids.

As a direct consequence of their dual basic-acidic nature, amino acids can react with both acids and alkalis. Such behaviour is termed *amphoteric*. At a certain pH the amino acid will be ionized with a negative carboxyl group $(-COO^-)$ balanced by a basic amino group $(-NH_3^+)$. This pH is called the *iso-electric point* and the amino acid is called a *zwitterion*. If acid is added and the pH falls below the iso-electric point the amino acid becomes positively charged, and conversely becomes negatively charged as the pH is raised.

Figure 1.39 Iso-electric point of an amino acid

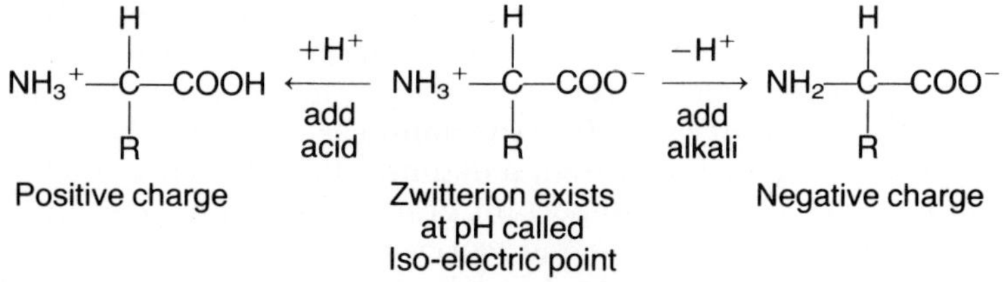

This phenomenon can be useful if the amino acids are placed in an electric field as positive amino acids will migrate towards a negative electrode and negative amino acids towards a positive electrode. This is the process of *electrophoresis* which is used to separate amino acids or proteins.

Amino acids can combine through their amino and carboxyl groups. When two amino acids condense together in this way a *dipeptide* is formed and where many combine a *polypeptide* is formed and finally a protein. The link is called a *peptide bond* and is shown in Figure 1.40.

Figure 1.40 Peptide bond

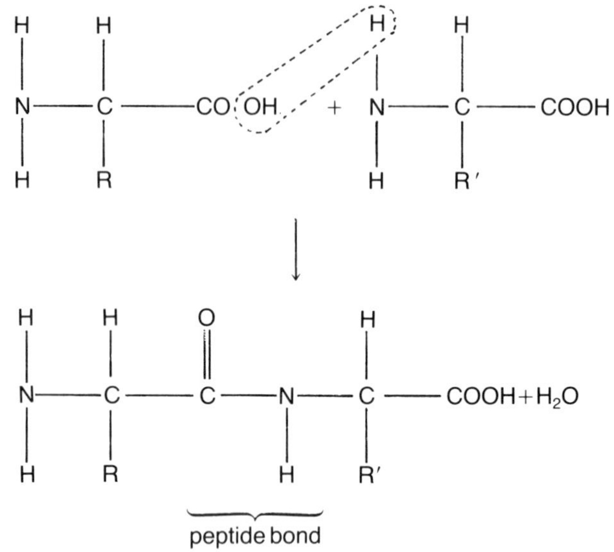

peptide bond

Each amino acid is linked to the next by peptide bonds to build up a chain of hundreds or even thousands of amino acids. As only about twenty amino acids are involved they are linked in an infinite variety of combinations to make an enormous number of different proteins.

1.4.1 The structure of proteins

The structure of proteins is different from other food components in that several orders of complexity have been recognised. Firstly, there is the *primary structure* which is the sequence of amino acids in the protein chain. In the *secondary structure* the amino acids are further linked by various bonds to give the protein a definite shape which is often in the form of a spiral. The most important group involved in cross linking is the $-SH$ group found in the amino acid cysteine. This forms a *disulphide* bridge. This bridge is important in bread-making as it gives elasticity and extensibility to the dough.

Other links are formed between the amino acids which contribute to the coiling of the chain of acids. These include the ubiquitous hydrogen bonds and the electrostatic attraction between positively charged amino groups (NH_3^+) and negatively charged carboxyl groups (COO^-).

Figure 1.41 Disulphide bridge

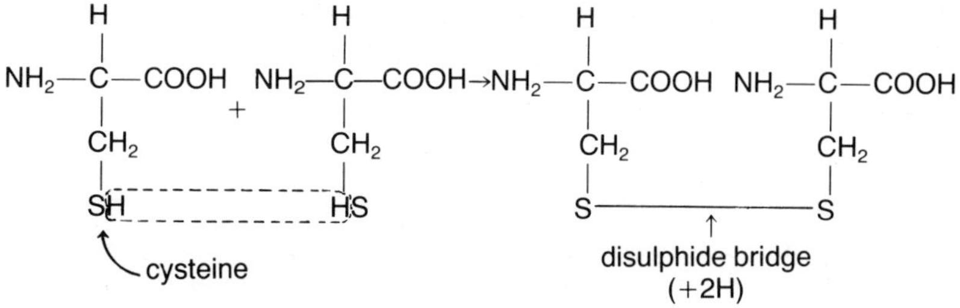

$$NH_2-\underset{\underset{\underset{SH}{|}}{\underset{CH_2}{|}}}{\overset{\overset{H}{|}}{C}}-COOH \quad + \quad NH_2-\underset{\underset{\underset{HS}{|}}{\underset{CH_2}{|}}}{\overset{\overset{H}{|}}{C}}-COOH \rightarrow NH_2-\underset{\underset{\underset{S}{|}}{\underset{CH_2}{|}}}{\overset{\overset{H}{|}}{C}}-COOH \quad NH_2-\underset{\underset{\underset{S}{|}}{\underset{CH_2}{|}}}{\overset{\overset{H}{|}}{C}}-COOH$$

cysteine

disulphide bridge
(+2H)

Figure 1.42 Secondary protein structure (diagrammatic representation showing cross links)

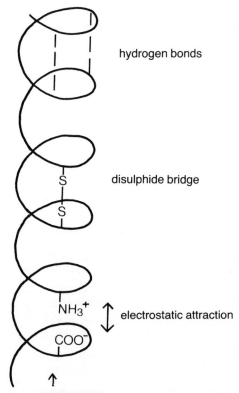

hydrogen bonds

disulphide bridge

electrostatic attraction

chain of hundreds of amino acids

This spiral or helix of the secondary structure may be folded over and held firmly by cross-links to form a globule, this is the *tertiary* structure.

49

There are even further protein structures such as fibres. Water plays an important role in these structures as without it they uncoil and lose that particular structure to varying degrees.

The structure of a protein may be readily changed by a number of agents causing molecules to aggregate and precipitate. This process is known as *denaturation,* and although usually it is irreversible, it can on occasions be reversible. Denaturation can be caused by heat, acids, alkalis, heavy metals, salt and ethanol. Violent agitation can also cause denaturation.

Reversible denaturation of a protein is slight unwinding of the polypeptide chain caused by mild denaturing conditions. If the protein is removed from the conditions it can regain its original structure and properties. This has been found recently in enzymes (see Section 1.4.2.2) which are sometimes only reversibly denatured in blanching and reactivate themselves during storage of the food.

Unfolding of the molecule occurs in *irreversible denaturation* leading to loss of some properties of the protein, but often making it more digestible. This is the reason for cooking of meat and many protein foods. The cooking of an egg illustrates the irreversible changes which occur in a protein when it is denatured. Solubility of the protein is lost; viscosity increases enormously; the egg becomes opaque; but digestibility is improved.

1.4.2 Classification of proteins

The simplest way of classifying proteins is by their function. There are three main functions for proteins – structural, physiologically active and nutrient.

Table 1.8 Classification of proteins by function

Function	Examples
Structural	cellular membranes muscle skin
Physiologically active	enzymes hormones (not all) blood proteins nucleoproteins
Nutrient	meat proteins – supply all essential amino acids

1.4.2.1 Structural proteins

Structural proteins possess mechanical strength which is due to an organised arrangement of their component amino acid and polypeptide chains. Some structural proteins are quite rigid whereas others are elastic. The inelastic or rigid proteins have polypeptide chains packed closely together and cross-linked with hydrogen bonds. This tends to make 'sheets' of protein much like a sheet of silk.

Figure 1.43 Sheet structure of a protein

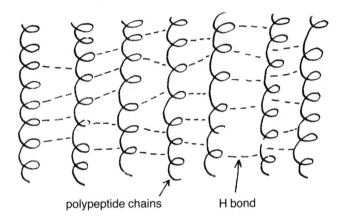

polypeptide chains   H bond

Elastic and extensible proteins are in the form of coils, or 'spring-like' as they will stretch but then revert to their original shape.

The coils are held together by hydrogen bonds and by disulphide bridges $(-S-S-)$. In wheat gluten there is an equilibrium between the $-SH$ groups and the disulphide bridges thus:

$$-SH + HS- \rightleftharpoons -S-S- + 2H$$

As fermentation proceeds by the yeast in bread making the disulphide bridges split to sulphydryl groups $(-SH)$ but then reform to make disulphide bridges. Thus the dough is extensible but retains its elastic properties due to the disulphide bridges. Muscle myosin, skin epidermis and blood fibrinogen are similar elastic proteins.

1.4.2.2 Physiologically active proteins

Enzymes are the largest single group of proteins and are often referred to as organic catalysts. In fact without enzymes no biological process could take place and life would be impossible.

Enzymes are very specific, some acting on one particular bond in one organic substance. The substance upon which an enzyme acts is called the *substrate*. Usually enzymes are named after the substrate upon which they act:

enzyme name = substrate + ase
  eg maltase = malt (ose) + ase
  pectinase* = pectin + ase
  (a general name for a group of enzymes)

Like proteins enzymes can be classified in a number of ways but the most common method is now by the type of reaction they catalyse.

Table 1.9 Types of enzymes

Enzyme	Reaction performed
Oxidases	Oxidation – add oxygen or remove hydrogen
Reductases	Reduction – add hydrogen or remove oxygen
Transferases	Transfer groups between molecules
Isomerases	Rearrange molecules into different structures
Synthetases	Build up more complex molecules from simpler ones
Hydrolases	Hydrolysis – adding water

Enzymes have the effect of reducing a single-step high energy reaction into a multi-stage process involving small amounts of energy. They are, therefore, highly selective and the reaction between an enzyme and its substrate has been likened to a lock and key. Figure 1.44 shows diagrammatically what is thought to happen in an enzyme catalysed reaction involving splitting a substance into two smaller units.

Figure 1.44 Enzyme action

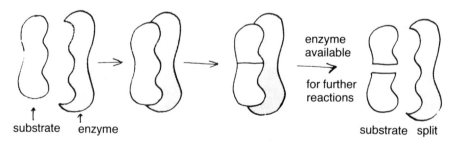

substrate   enzyme   substrate split

Enzymes are surface catalysts and are therefore readily blocked by a large variety of substances which can adhere to their surface. Enzymes often need the help of another non-protein substance called a *co-enzyme* to make a reaction possible. Often these co-enzymes are vitamins or derived from vitamins. In a similar manner enzymes sometimes need *activators* which are normally metals such as magnesium and occasionally non-metals such as chlorine (as chloride).

Enzymes prepared from a wide range of sources have many uses in food manufacture. Invertase is used extensively in confectionery to produce invert sugar in a wide range of products. Meat has been tenderized by enzymes such as papain from paw-paw or papaya, bromelin from pineapples or ficin from figs. Pectinases from fungal sources have been used to remove pectin hazes from wines. In foods which have not been heat-treated, enzymes may be active during storage and cause significant deterioration. Even at frozen temperatures some enzymes may be active and produce flavour, colour or texture changes in the product. Before canning, drying or freezing a number of foods, particularly vegetables, the product is *blanched* to inactivate the enzymes. The enzyme peroxidase is one of the most heat-resistant enzymes and it is assumed that testing for the absence of peroxidase after blanching will indicate the inactivation of all enzymes present in the food. However, recent work has shown that although there are some more heat-resistant enzymes, a number of enzymes can actually reactivate themselves during storage and damage the product. Blanching involves a short heat treatment of one or two minutes with boiling water or steam.

Figure 1.45 Peroxidase test

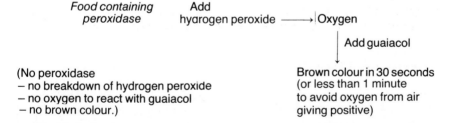

Some *hormones* are also proteins and hold key positions in many metabolic processes. *Nucleoproteins* are important physiologically active proteins found in food, particularly in meat. They are conjugates of nucleic acid and protein. There are two classes of these proteins: deoxyribonucleic acid (DNA) proteins and ribonucleic acid (RNA) proteins. There are a number of different blood proteins but haemoglobin is important because of its function as an oxygen carrier.

The red colour of meat is mostly due to the pigment myoglobin which has a structure similar to haemoglobin. Haemoglobin is lost from meat when the blood is drained from the carcass after slaughter and so does not contribute to the colour of meat significantly. (See Section 2.2.1.)

1.4.2.3 Nutrient proteins

As the body is unable to synthesise the essential amino acids it is necessary for certain nutrient proteins to be taken in the diet. Animal proteins are the obvious main individual nutrient proteins but plant proteins are now consumed in greater quantities. The high cost of animal proteins has led to the development of a wide range of *novel proteins* or *meat substitutes*.

1.4.3 Novel proteins (meat substitutes or analogues)

New types of proteins have been developed because of the high cost and inefficient production of protein in animals. Many parts of the world are suffering from protein malnutrition or 'kwashiorkor'. It is hoped that new proteins will help to reduce this problem.

Novel proteins can be manufactured to have a protein quality of high biological value, similar to meat, or they can enrich poor diets based on cereals or crops such as cassava.

Two general types of novel proteins are being produced and new ones are being developed, based on naturally occurring proteins such as soya protein, or biosynthesised protein produced by fermentation of various substrates such as waste carbohydrate.

Soya protein

Yet again we come across the useful soya bean plant. It is used to make *textured vegetable protein* or TVP. Soya bean meal is defatted and then mixed with water. The dough-like mass is then extruded and at the same time heated under pressure. It dries into a mass of spongy-like consistency, which can be cut into chunks, or ground into granules. Before the extrusion stage, colour, flavouring and salt are often added. The dried pieces of TVP are usually referred to in 'meat terms' such as mince, and chunks. This product probably would have been more successful if this obvious comparison with meat was not made, as TVP invariably loses in any such comparison. The dried TVP pieces must be reconstituted in water before use or in a suitably flavoured sauce.

Spun soya is an improvement of TVP as it tends to duplicate the more fibrous texture of meat products. Soya proteins do have some disadvantages. Soya proteins are deficient in the essential amino acid methionine and therefore need fortification with this if used as primary protein sources. Some people have difficulty in digesting completely soya protein and others are affected by carbohydrates present, such as raffinose, which cause excessive flatulence.

Biosynthesised proteins (or single cell proteins)

These proteins are produced by bacteria or fungi growing on suitable substrates usually by a continuous fermentation method. Waste carbohydrate material can be converted by some fungi to proteins; this protein is then extracted and used as a food supplement. The proteins, themselves, taste like mushrooms and are generally in a powder form after extraction, and are suitable for addition to other foods.

Consumer resistance to this type of product must be overcome before they are readily acceptable. Toxicological studies have to be carefully carried out and nutrition trials have to be extensive. Starving people surprisingly often show considerable resistance to new products of this nature.

1.4.4 Browning reactions

There are many occasions when it is desirable to have browning of foods particularly in cooking by roasting, baking and grilling. There are equally numerous occasions when a brown product indicates deterioration and loss of nutritive value. There are two main groups of browning reactions, those involving enzymes and those occurring between proteins and carbohydrates.

1.4.4.1 Enzymic browning

When many varieties of apple are eaten or peeled they rapidly start to develop a brown colouration. Many other fruits and vegetables show this

Figure 1.46 Catechol

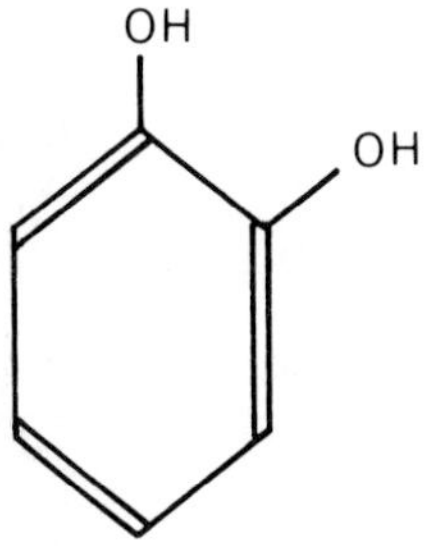

(NB The best substrates for the polyphenolases must have two adjacent–OH groups on a benzene ring.)

browning, particularly potatoes, pears, avocados and bananas. Browning occurs when the tissue is physically damaged or is diseased.

The enzymes which cause this browning are called *polyphenolases* or polyphenol oxidases. The substrates for the enzymes are phenolic compounds, particularly diphenols such as catechol (in apples).

Monophenols can also act as substrates but are much slower in their reaction. Some fruit and vegetables possess the enzymes and not the substrates, and therefore do not brown, eg citrus fruits. The third requirement for a fruit to brown is oxygen or air. A series of reactions occurs after the initial action of the enzyme on the substrate eventually to produce brown pigments or *melanoidins*.

Figure 1.47 Enzymic browning

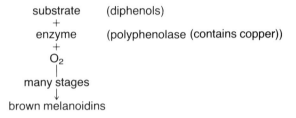

This browning reaction is undesirable and unsightly, particularly in fruit salads, peeled apples and potatoes. Cider is a golden brown colour because of this reaction and is probably the only occasion when the reaction is desirable.

There are many methods of preventing the reaction, but only a few are used. The enzymes can be destroyed by heat or inactivated with sulphites or sulphur dioxide. Oxygen can easily be excluded from the food, for example peeled potatoes are immersed in water. The water should be boiled, to drive out any dissolved oxygen, and then cooled before use. Addition of a little salt will help. Acids will prevent the action of the enzymes and lemon juice is often used in cookery for this purpose.

1.4.4.2 Non-enzymic browning

When foods are roasted, baked or grilled, non-enzymic browning occurs and is considered desirable because of the appetizing colour, odour and flavour it produces. Sometimes, however, it is undesirable and indicates deterioration in a food; for example, the gradual browning of dried milk powder.

Maillard in 1912 noticed that brown colouration was produced when a solution of glucose was heated with the amino acid glycine. This reaction between the amino group (NH_2) of a protein or amino acid and the (potential) aldehyde group of a reducing sugar is called the *Maillard reaction*.

Figure 1.48 Maillard reaction

$$-NH_2 + -C\,H\,O$$

(from amino acid or protein)   (from a reducing sugar)

|

number of rearrangements

↓

brown melanoidins

(*NB* The reacting groups are attached to their relevant molecules and in some cases large proteins are involved.)

There will be some loss in nutritive value of proteins involved in non-enzymic browning. Amino acids containing extra amino groups, for example lysine, are obviously more susceptible. The complexity of foods poses many problems for controlling this type of browning. Products which brown slowly during storage should be kept at a lower temperature. The moisture content of a dried product is critical. Skimmed milk powder requires about 5% moisture before browning can proceed. The Maillard reaction proceeds more rapidly under alkaline conditions and therefore addition of an edible acid such as citric acid will slow down the reaction. Some foods can be treated with enzymes to remove either the protein or the sugar, eg in egg white, glucose is removed by yeast fermentation prior to drying.

Sugars can undergo browning in the absence of an amino group and this is *caramelization* (often confused with the Maillard reaction). Caramelization occurs when sugars are heated above their melting points to produce a range of brown substances collectively known as caramel. Ascorbic acid, vitamin C, can undergo a form of non-enzymic browning in fruit juices, particularly in grapefruit, when it is oxidised and eventually forms brown pigments.

Review

1. Proteins – largest known molecules, built up of amino acids

2. Amino acids – always contain an amino group (NH_2) and carboxyl group (COOH)

3. Amino acids – neutral – eg glycine
   basic – eg lysine
   acidic – eg glutamic acid

   – neutral amino acids: $1 \times NH_2$, $1 \times COOH$
   basic amino acids: $2 \times NH_2$, $1 \times COOH$
   acidic amino acids: $1 \times NH_2$, $2 \times COOH$

4. Eight essential amino acids:
   valine, leucine, isoleucine, phenylalanine,
   threonine, methionine, tryptophan, lysine
   Children also require: histidine and arginine

5. Amphoteric – amino acids can act as alkalis or acids

6. Iso-electric point – amino group ionised to NH_3^+
   (occurs at certain pH) balanced by COO^-

7. At iso-electric point amino acid called a zwitterion

8. Amino acid joined to another by peptide bond

9. Primary structure of a protein – sequence of amino acids in the chain

10. Secondary structure – definite shape, usually a helix caused by cross-linking by various bonds

11. Tertiary structure – secondary structure folded over and held by cross-links to form globule

12. Structural proteins – cellular membranes – skin, muscle

13. Physiologically active proteins – enzymes, blood proteins

14. Nutrient proteins – meat proteins – supply all essential amino acids

15. Enzymes – largest group of proteins

16. Enzyme name = substrate + ase

17. Classification of enzymes by nature of reaction catalysed –
    eg oxidases – perform oxidation
    isomerases – rearrange molecules into different structures

18. Enzyme action is very specific – compared with a lock and key

19. Non-proteins may be needed by enzyme to make reaction possible – known as co-enzymes

20. Browning reactions – enzymic – requires polyphenolases+
diphenol substrate+oxygen
– non-enzymic – require reducing sugar+amino group
(in protein or amino acid)

21. Maillard reaction – non-enzymic browning reactions
to produce brown melanoidins

22. Caramelization – browning of sugars when heated –
not involving an amino group

Practical exercises:
Proteins and browning reactions

Try the following tests on dried egg, dried milk and gelatine.

1. Test for all proteins – Biuret test

Take a test-tube about ⅛th full of protein solution (or suspension) and add an equal quantity of sodium hydroxide solution. Add one drop of copper sulphate solution. A violet colour is obtained for all proteins.

2. Heating proteins

Take a sample of protein and strongly heat in a tube (ignition tube). Test the end of the tube with moist red litmus paper. A blue colour is produced as fumes of ammonia are given off because of the nitrogen content of the protein.

3. Protein denaturation

Add some egg albumin to water in a test tube and pour a few cm^3 in other tubes for the following tests:—
  (a) heat gently, then boil
  (b) add a few drops of hydrochloric acid
  (c) add a few drops of sodium hydroxide
  (d) add a few drops of mercuric chloride solution (**poison**)
Observe the effects on the protein solution.

4. Millon's test – for tyrosine

Take about ⅛th of a test-tube of protein solution and add a similar amount of Millon's reagent (mercurous and mercuric nitrates in nitric acid – **poisonous**). A white precipitate is formed which becomes brick-red when heated in a water-bath.

5. Sakaguchi's test – for arginine

To a small quantity of protein solution add a few cm^3 of sodium hydroxide solution. Add about 4 drops of a 2% solution of α-naphthol (in alcohol), a drop of sodium hypochlorite solution (or bleaching powder). Arginine causes the production of a carmine colour.

6. Xanthoproteic test – for amino acids with a phenyl group

To a small quantity of protein solution add a few drops of concentrated nitric acid (CARE!). Boil in a water bath. Cool under a running tap and add ammonium hydroxide to make it alkaline. An orange colour is produced by phenylalanine, tyrosine and tryptophan.

7. Enzymic browning reactions

Slice an apple into a number of thin slices and carry out the following:
  (a)   break one slice
  (b)   crush one slice
  (c)   immerse one slice in water
  (d)   immerse one slice in water which has been boiled then cooled
  (e)   immerse one slice in a dilute acid solution (eg citric acid solution)
  (f)   immerse one slice in a 5% salt solution
Repeat for potato, and observe for any browning produced.

If available, treat slices with catechol, resorcinol, and tyrosine.

8. Non-enzymic browning – the Maillard reaction

Make a solution of glucose (about 10%), add some protein or amino acid, as available. Boil in a water bath and note any browning.
Repeat, but:
  (a)   add a few drops of acid
  (b)   add a few drops of sodium hydroxide
  (c)   add a few drops of a sulphite solution

1.5 Minerals

The human body requires a range of mineral substances which can be absorbed from food. Unfortunately in many cases food processing and cooking techniques reduce the level of available minerals in foods. This is sometimes offset by the enrichment of foods, for example, breakfast cereals, with a number of minerals particularly iron. Some of the minerals are only required in trace amounts and excessive intake of them may cause poisoning.

Calcium and phosphorus

These two elements will be considered together as they exist in the body as calcium phosphate which makes up the structure of bones and teeth. There is about 1–1·5 kg of calcium and 0·75–1·0 kg of phosphorus in the body. Fortunately phosphorus is a natural constituent of plant and animal cells and, therefore, is readily obtained from food. Shortage of calcium in the diet, however, affects the bones and teeth which can become weakened and soft. Only about 40% of calcium in food is absorbed by the body, so for some people there is a risk of calcium deficiency. This deficiency leads to the disease of osteoporosis. The main sources of calcium are dairy products, especially cheese, bread and some vegetables. Some vegetables and wholemeal flour contain a substance called *phytic acid* which can combine with calcium and make it unavailable to the body. Similarly *oxalic acid*, poisonous in higher quantities, will form calcium oxalate which is unavailable to the body. This occurs in spinach, for example.

Iron

Iron is an essential constituent of the blood pigment, haemoglobin, and is involved in transport of oxygen around the body. Anaemia is a clinical condition associated with iron abnormalities in metabolism. However, deficiency of nutrients other than iron may be the true cause. The daily intake for an adult male should be about 10 mg; females require slightly more, particularly during menstruation, pregnancy and lactation.

Fortunately, the iron used in haemoglobin is re-cycled when red blood compounds are broken down in the body after their usefulness has ended. A wide range of foodstuffs contain iron particularly liver, some fish, kidney, eggs, brown bread and flour. A number of foodstuffs are enriched with iron, for example, breakfast cereals have iron added at the rate of 6–10 mg/100 g.

There is evidence that some forms of iron used for food enrichment purposes are poorly utilized by the body. Iron must be in the iron II or ferrous form to be utilized. The oxidised iron III or ferric form is, in fact, more common in foods than the reduced iron II form, and must be

61

reduced to iron II in the body. The absence of oxygen, or reducing conditions, will ensure that the iron II form is taken into the body. Ascorbic acid or vitamin C is a reducing agent and will aid the intake of iron in this manner. A glass of orange juice, particularly a formulated juice with added ascorbic acid, will ensure the utilization of iron from an enriched breakfast cereal which is eaten after the juice.

In food processing loss of iron occurs as it dissolves into processing water which may be discarded. Finely chopped or diced vegetables present a large surface area to water and therefore many substances including iron readily dissolve in the water. In certain vegetables iron accumulates near the peel and therefore, peeling will usually remove the iron.

Potassium and sodium

Both these elements are important in cell and body fluids. Potassium is used in soft tissues and helps to control pH and osmotic pressure within the cell. Sodium and potassium usually occur with chlorine and are taken into the body from a wide range of foods and chlorides. Sodium and potassium are involved in maintaining an electrolyte balance in the body. Excessive amounts are excreted by the kidneys so that a balance particularly between the concentration of sodium and chloride is maintained.

There is growing evidence that many people eat too much salt (sodium chloride) with their food. Not only is salt used in cooking but it is liberally added to the food during the meal. This excess consumption of salt leads to excess sodium in the body and the electrolytic balance is upset. Excessive salt has been shown to lead to high blood pressure, some stress and nervous symptoms and possibly to kidney disease. People suffering from high blood pressure are usually put on a salt-free diet and are recommended salt substitutes, particularly potassium chloride. High salt intake also causes the retention of body fluid, which obviously causes weight increase and has been implicated with the symptoms of premenstrual tension in some women.

The body can obtain enough sodium from a normal diet without the unnecessary additions in the food factory or kitchen.

The current theory is that salt will only cause high blood pressure in those with a family history of such problems.

Trace elements

Very small amounts of some minerals are needed by the body. The essential trace elements are: chromium, cobalt, copper, fluorine, iodine, manganese, molybdenum, nickel, selenium, silicon, tin, vanadium and zinc. Many of these substances are essential constituents of enzyme systems but some are involved in other functions. For example, iodine is essential for the thyroid gland to produce the hormone thyroxine. The body requires about 20–40 mg of iodine daily and normally this is available from fish, other seafoods and vegetables. Shortage of iodine can cause goitre, which is an enormous enlargement of the thyroid gland (known as 'Derbyshire neck'). Salt is 'iodized' with potassium iodide to supplement the iodine from food. However, certain processes lower iodide levels in food and supplementation of salt may be insufficient particularly for people on low salt diets.

Review

1. Calcium most abundant mineral in body, followed by phosphorus

2. Calcium – from dairy products and some vegetables
   – bound by phytic or oxalic acid

3. Phosphorus – occurs in all cells – readily found, therefore, in foods

4. Iron – essential for haemoglobin
   – must be absorbed in iron II or ferrous form

5. Iron used in haemoglobin can be recycled in body

6. Potassium and sodium involved in electrolyte balance in body and control of osmotic pressure within cells

7. Excess salt consumption, can cause high blood pressure and fluid retention in the body

8. Trace elements – very small amounts required generally by enzyme system

9. Lack of iodine causes goitre – 'Derbyshire neck'

10. Higher amounts of trace elements are toxic

Practical exercises: *Minerals*

A food sample should be converted into ash before the following tests are performed. The sample is placed in a crucible which is heated over a bunsen flame to destroy the organic matter. The crucible is then heated in a muffle furnace for 2 or 3 hours to ash the sample. (If a furnace is not available heat for a longer period with a bunsen, which for some products might be sufficient.)

To a few drops of a solution of the ash carry out the following:—

1. Chloride –

add a few drops of silver nitrate solution which gives a white precipitate with chloride.

2. Sulphate –

add a few drops of barium chloride solution. A white precipitate, which does not dissolve on adding hydrochloric acid, indicates the presence of sulphate.

3. Carbonate –

add a few drops of dilute sulphuric or hydrochloric acid. Effervescence will occur if a carbonate is present. Also test the dry ash with acid.

4. Phosphate –

add concentrated nitric acid (CARE!) and then ammonium molybdate solution (10%). Heat in boiling water and look for the formation of a yellow colour or precipitate.

5. Flame tests for metals –

Use a special wire loop and dip into the ash solution. Place the loop in a bunsen flame and note the colour:

       potassium – lilac
       calcium   – red
       copper   – green
       sodium   – yellow

6. Test for iron (Iron III or ferric compounds)

To a few drops of the ash solution add hydrochloric acid and then ammonium thiocyanate solution (10%). A red colour indicates the presence of ferric (iron III).

Iron II or ferrous compounds can be detected by treating the sample with hydrogen peroxide and gently heating. This converts iron II to iron III, which is then detected as above.

1.6 Vitamins

Vitamins are organic substances which are required in small quantities but cannot be synthesised by the body. A substance which acts as a vitamin for one animal may not always be required by others, for example, many animals synthesise their own vitamin C. Low levels in the diet of a vitamin will cause a *vitamin deficiency disease,* which may be fatal if not remedied. However, although figures may be quoted for the vitamin content of a particular food there is no guarantee that the vitamin can be absorbed from the food into the body during digestion. Many vitamins can be bound with other substances in the food. A balanced diet will ensure that the body receives at least its minimum requirement of vitamins. High levels of certain vitamins have been advocated by some experts, and some fanatics, to cure all manner of illnesses. The body can become adjusted to higher levels of vitamins and can develop deficiency diseases when the level of vitamin intake is lowered to a normal level.

An argument often made against processed foods is that they contain fewer or lower concentrations of vitamins or even none at all. Certainly processing reduces the level of many vitamins, but equally bad storage and poor cooking techniques take their toll of vitamins in the diet. Often the food manufacturer has the opportunity of adding vitamins to a food product, thus making it a better source than the original 'fresh' product. Margarine, by law, must contain 900 µg/100 g of vitamin A and 8 µg/100 g of vitamin D, which is similar to the vitamin level of summer butter. However, most butter often falls below these levels!

It is normal to divide the vitamins into two groups: water-soluble and fat-soluble vitamins. There has been some confusion over the naming of vitamins and the letter each one was assigned. Groups of vitamins, for example the B group, have similar structures and therefore have been given the same letter but with a number postscript. To avoid difficulties it is better to remember the names of the vitamins in common usage. The vitamins are listed in Table 1.10 with their main sources and uses in the body.

1.6.1 Water-soluble vitamins

1.6.1.1 Ascorbic acid (Vitamin C)

More has been written about this vitamin than any other and probably more has been claimed for its benefits. There is some evidence that it helps to speed up the curing of a common cold and it may also help to alleviate the symptoms of a number of other illnesses, some believe it helps prevent cancer. This is based on the evidence that animals which synthesise their own vitamin C suffer such diseases less often. The

Table 1.10 Vitamins

Name	Letter	Source	Use in body
Water-soluble			
Ascorbic acid	C	Potatoes, fruit, rose hips	Proper formation of teeth, gums, blood vessels
'B complex':—			
Thiamin	B_1	Wheatgerm, yeast	
Riboflavin	B_2	Kidney, liver, cheese	
Pyridoxine	B_6	Yeast, liver, grain	
Cyanocobalamin	B_{12}	Liver, fatty fish	Co-enzymes in many reactions
Nicotinic acid	—	Yeast, meat, liver	
Pantothenic acid	—	Many foods	
Biotin	—	Liver, yeast	
Folic acid	—	Broccoli, watercress, liver	
Fat-soluble			
Retinol	A	Fish oils, milk, green vegetables	Necessary for skin, normal growth and eyesight
Calciferols	D	Butter, margarine, fish oils, fatty fish	Necessary for calcium absorption in formation of bones and teeth
Tocopherols	E	Wheatgerm, vegetable oils	Important in cell metabolism
Naphtho-quinones	K	Spinach, kale, cauliflower	Involved in blood clotting

function of ascorbic acid in the body is still not fully understood, but it is involved in the formation of hydroxyproline which is a key constituent of collagen. Deficiency of vitamin C leads to the disease of scurvy which used to be the cause of misery and finally death for sailors on long voyages up to the last century. Symptoms of the disease include lassitude, swelling and bleeding of gums, loosening of teeth, bruising, internal bleeding and finally death. Fresh fruit, from any source, quickly cured the disease. British sailing ships during the last century put into ports of the West Indies to take on board lime juice which was known to cure the sickness, hence the term for British sailors of 'Limies'.

Vitamin C is ascorbic acid in either the oxidised or reduced form. The latter, is a powerful reducing agent and use can be made of this property in food processing.

Figure 1.49 Vitamin C

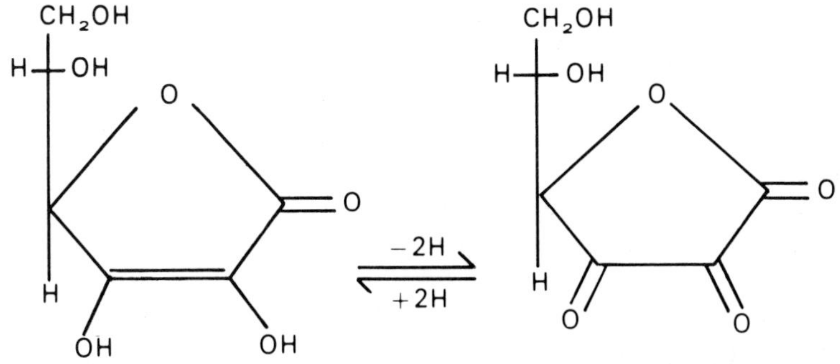

ascorbic acid (reduced form) ⇌ dehydroascorbic acid (oxidised form)

Some fruits, particularly berries, are rich in vitamin C, for example blackcurrants can contain 200 mg/100 g, rose hips 175 mg/100 g and strawberries 60 mg/100 g. Oranges, considered by many a good source, contain 50 mg/100 g. Tropical fruits often contain considerable amounts of the vitamin and similarly fruit exposed to the sun usually contains more than fruit in the shade. Outer layers of many fruits contain more than inner layers. New potatoes contain perhaps 30 mg/100 g which falls to as little as 8 mg/100 g after 6 months storage, but nevertheless, because of the large quantities eaten, potatoes have contributed significant amounts to the diet.

Although figures are quoted for levels of the vitamin in a certain food these levels are frequently much lower as the vitamin is easily destroyed by oxidation, heat and water extraction. When a foodstuff is processed it is often cut, diced or chopped and because of this enzymes are released which catalyse the oxidation of ascorbic acid. Acid will greatly retard the loss due to oxidation, as oxidation is most rapid under alkaline conditions and in the presence of small amounts of copper, such as copper pans.

When cooking some vegetables, such as runner beans, it is a common practice to add a small amount of bicarbonate of soda (sodium hydrogen carbonate) in order to prevent chlorophyll loss, thus keeping the bright green colour of the fresh vegetable. Chlorophyll is stable under alkaline conditions which is directly opposite to ascorbic acid. Thus the addition of bicarbonate of soda, making the vegetable slightly alkaline, will cause the loss of vitamin C. However, does this matter in a balanced diet? Most people do not rely on runner beans as a source of vitamin C and the benefit of cooking an attractive product instead of a muddy-grey one far outweighs this slight loss of vitamin C in the diet.

During cooking and processing losses of up to 75% of ascorbic acid may occur. Short processes involving small volumes of water incur the smallest losses. However, most processes cause significant losses, particularly dehydration, canning and freezing. Fortunately, if blanching is required in a process, for example before canning or freezing, ascorbic acid may be added to the blanching water. This not only increases the vitamin C level of the product but often produces colour and flavour improvements.

There is still debate as to what the daily intake of vitamin C should be. In 1969 the DHSS recommended 30 mg per day; in 1973 the US Food and Drug Administration recommended 60 mg; and now some authorities say the figure should be as high as 100 mg. However, an intake of as little as 10 mg will prevent the symptoms of scurvy.

1.6.1.2 The vitamin B complex

Often vitamins of this group are found associated with protein, for example in liver, kidney, cheese and yeast. Many of the vitamins in the group act as co-enzymes and therefore deficiency in the diet may interfere with vital enzyme controlled reactions in the body.

Thiamin (Vitamin B_1)

This vitamin follows closely the behaviour of ascorbic acid in foods, although normally it is not found with ascorbic acid. It is readily soluble in water and is rapidly destroyed by heat in neutral or alkaline solutions, but is relatively stable in acid conditions. The vitamin will leach out of a food in proportion to the amount of water in contact with it, the surface area of the food exposed, and the degree of agitation of the food in the water. Any process that minimises the length of time a food is subjected to these three factors, the less will be the vitamin loss. In cooking, thiamin loss may vary from 15–60% depending on the amount of water used and the temperature of cooking.

Serious losses of thiamin occur when sulphur dioxide or sulphites, such as sodium metabisulphite, are used as preservatives. These sulphur compounds are also used to inhibit browning, for example in frozen chipped potatoes. In such products thiamin is readily broken down.

The disease known as beri-beri, in any of three different forms, occurs when the diet is deficient in the vitamin. In the Far East polishing of rice, by removing all the outer layers of the grain, caused widespread outbreaks of the disease. If the rice is par-boiled before milling, then dried, the vitamin migrates from the outer layers of the rice grain into the centre and is therefore not lost during milling. The disease can show a range of symptoms developing through loss of appetite, muscular weakness, palpitations, fever, and sudden heart failure. Treatment with thiamin causes rapid recovery.

In Western diets the disease is rare but has been found in people suffering from anorexia and alcoholism. An average man requires about 1·5 mg per day of thiamin.

Riboflavin (Vitamin B_2)

Although classed as a water-soluble vitamin it is not as soluble as others in the group. In most foods the vitamin is heat-stable and can withstand boiling in acid conditions. However, like ascorbic acid and thiamin it is decomposed by heat under alkaline conditions, which fortunately rarely occur in foods. Again the best sources of the vitamin are liver, yeast and dairy products, but it is fairly common in most foods. Deficiency of the vitamin shows itself in skin problems, particularly cracking of the skin around the mouth. About 2 mg per day is needed in the diet. Riboflavin and its derivative riboflavin phosphate are useful food colours giving attractive yellow shades.

Pyridoxine (Vitamin B_6)

Deficiency of this vitamin is rare as it occurs in a fairly wide range of foods. Like thiamin, however, it can be lost during the milling of grain. It can be destroyed by heat, particularly under alkaline conditions, and is susceptible to light. Dermatitis, nervous problems and some types of fits may be caused by deficiency of the vitamin. About 1–2 mg per day is sufficient for most adults.

Cyanocobalamin (Vitamin B_{12})

This vitamin, which has an extremely complex structure, is not found in fruits or vegetables. The main source is animal food, but it sometimes occurs in seaweed products and fungi. Pernicious anaemia results if the diet is deficient in the vitamin and usually strict vegetarians (vegans), who eat no animal food, are the only people susceptible. The recommended daily amount required in the diet is only 3 µg (ie 0·000003 g). The vitamin is heat-stable and is not usually lost in processing.

Nicotinic acid (niacin)

The related compound nicotinamide has a similar biological function as a co-enzyme involved in the metabolism of carbohydrates. Readily soluble in water, this vitamin is unaffected, however, by most cooking and processing operations. Although found with other B vitamins in liver, yeast and meat, it can also be synthesised to a limited amount from the amino acid tryptophan. Deficiency disease takes the form of diarrhoea, mental problems and skin disorders, and is generally known as *pellagra*. This is an example of a vitamin which is often not available in a food as it is bound to some other substances and cannot be released by enzymes found in the human digestive system. This occurs particularly in cereals, and people living for example on maize, may develop symptoms of pellagra. The normal dietary requirement is about 18 mg per day.

Pantothenic acid

This vitamin, although present in most foods, is destroyed by dry heat, acids and alkalis. It forms part of the co-enzyme known as co-enzyme A which is involved in metabolism of lipids and carbohydrates. Daily requirements are about 6 mg, but deficiency disease is highly unlikely to occur.

Biotin

Biotin is generally unaffected by processing, but a protein called *avidin* in egg white will combine with it, making it unavailable. The bacteria in the human gut synthesise the vitamin so deficiency is very rare.

Folic acid (folacin)

Folic acid is one of the few nutrients which may be deficient in the UK diet, particularly the diet of the elderly and pregnant women. The latter are normally prescribed iron tablets supplemented with vitamins which include folic acid for this reason. Liver, broccoli and watercress are food sources, but some may be produced by intestinal bacteria.

Folic acid deficiency causes a type of anaemia in which red blood cells become enlarged (megaloblastic anaemia). People, particularly pregnant women, living on poor diets, such as those based on cassava, rice or wheat, have shown the symptoms. The vitamin is involved in the synthesis of a number of compounds in the body including some amino acids. Heat processing can destroy the vitamin particularly in the presence of oxygen or air. About 0·4 mg per day is needed by an adult.

1.6.2 Fat-soluble vitamins

1.6.2.1 Retinol (Vitamin A)

This was the first fat-soluble vitamin to be discovered and it occurs only in animal products such as fish oils, dairy products and liver. However, many green vegetables and carrots contain carotenoid substances (see Sections 1.7.1 and 1.7.1.1) which can be converted partially or completely into retinol. The carotenes (α, β & γ) show vitamin A activity and are often called *pro-vitamin A*.

In Figure 1.50 the ring at the left of the structure is known as β-*ionone* ring. Carotenoids must have this ring in their structures in order to act

Figure 1.50 Structure of retinol

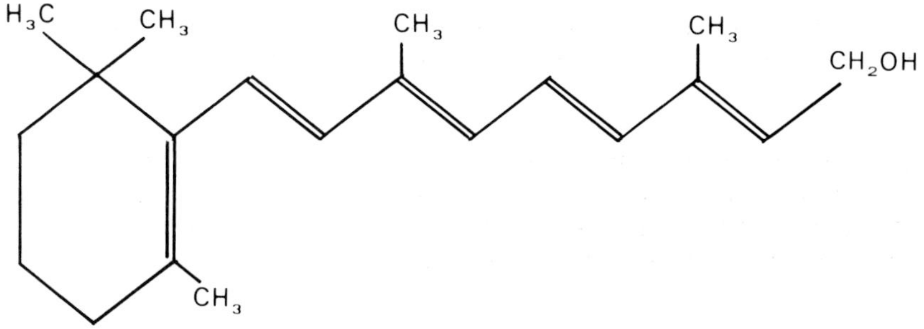

as a pro-vitamin A. Lycopene, the red pigment of tomatoes, has open rings at both ends of its structure and therefore shows no vitamin A activity. It is obvious, therefore, that β-carotene having two β-ionone rings can be converted into two molecules of retinol and therefore has double the pro-vitamin A activity compared with α- and γ-carotenes which have only one β-ionone ring. Unfortunately, carotenoids are not easily absorbed from the gut.

Vitamin A deficiency will cause a general deterioration in health and maintenance of skin. Colour and 'night blindness' are usually the first symptoms of deficiency. Pilots flying night missions in the War were encouraged to eat plenty of carrots to help them to see in the dark. Dairy products and fish liver oils, particularly, would have been a better recommendation!

It is possible to be poisoned by excessive intake of vitamin A which can cause drowsiness, skin and bone disease, and liver enlargement. Retinol is fairly stable during cooking but may be destroyed by oxidative rancidity. Dehydrated products, such as dried diced carrot, can quickly lose their colour (carotenoids) and the pro-vitamin A activity, if exposed to air or oxygen.

1.6.2.2 Calciferols (Vitamin D)

Vitamin D, not only occurs in fish liver oil and dairy products, but can be synthesised in the body by exposure to sunlight (UV light). By our definition of a vitamin, therefore, vitamin D should not be included. However, as we know, only too well, sunlight is often absent in the UK and people could show deficiency of the vitamin leading to the disease known as *rickets*. As calcium absorption and bone formation is impaired, particularly in children, the disease shows itself in the bending of bones in the legs and other deformities. Cod liver oil contains vitamin D type compounds and it is therefore a traditional remedy for rickets. Rickets has largely been eliminated with perhaps the exception of some immigrant communities. Some darker skinned immigrants, not eating foods containing sufficient calciferols, have shown symptoms of rickets. Ultra-violet light is not absorbed as much by a darker skin and therefore there is insufficient synthesis of the vitamin in the body.

Vitamin D is related structurally to cholesterol and is derived from this steroid in the skin.

Figure 1.51 Structure of calciferol

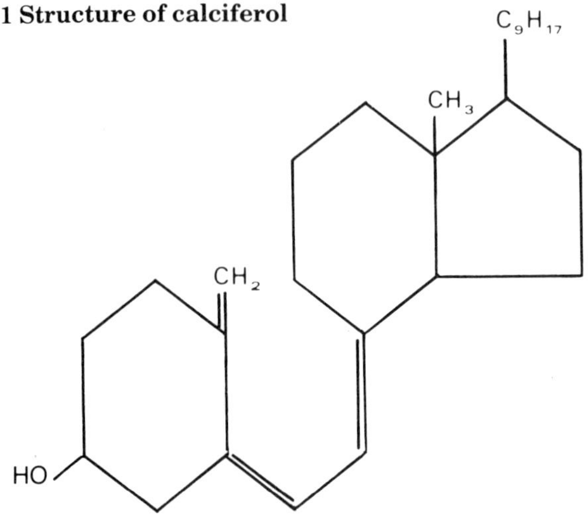

There are two calciferols which differ slightly. Cholecalciferol (vitamin D_3) is formed in the skin and ergocalciferol (vitamin D_2) can be prepared from yeast.

Vitamin D is unaffected by cooking processes and is resistant to heat. About 2·5 µg is sufficient for adults. Excessive intake can cause poisoning which damages the kidneys and may be fatal.

There is some evidence that Vitamin D prevents coronaries. Older people who are less able to synthesise Vitamin D are more susceptible to heart disease.

1.6.2.3 Tocopherols (Vitamin E)

There are eight different, but related, substances which show vitamin E activity. They occur particularly in wheat germ and in a number of vegetable oils, but are widely distributed in small amounts. α-Tocopherol is the most common form. Original research on the vitamin showed that deficiency produced sterility in rats. However, in humans dietary deficiency is unusual but may occur in babies, leading to anaemia and possibly blindness.

The role of vitamin E in human metabolism is still obscure but it may be involved in certain cell metabolic reactions. Many claims have been made for the health giving properties of vitamin E and its curing of diseases such as diabetes or heart disease. Large doses are said to have cosmetic uses and possible 'anti-ageing' effects. None of these claims has been vindicated.

However, vitamin E is a widespread naturally occurring antioxidant. It has clearly been shown to retard oxidative rancidity of lipids, particularly in vegetable oils. This antioxidant property will also be of value in protecting other vitamins, such as retinol, which are susceptible to oxidation in certain food products.

As tocopherols can be oxidised easily they are often lost in processing, for example, the vitamin E activity of wheatgerm is lost completely during the wheat milling process. Excessive intake of the vitamin is not toxic and therefore use could be made of added vitamin E to certain foods to protect against oxidative rancidity.

Vitamin E has been shown to protect against the damaging effects of free radicals (qv) on tissue membranes in the body.

1.6.2.4 Naphthoquinones (Vitamin K)

The fat-soluble vitamin is found in two forms: vitamin K_1 found in green vegetables and vitamin K_2 which is produced by bacterial decomposition of protein in the gut. Biochemists have produced a water-soluble version, K_3, which can be used to treat the deficiency symptoms. The main function of the vitamin is in blood clotting as a symptom of deficiency is uncontrolled bleeding after even a slight injury. Fat absorption in babies may be impaired by insufficient vitamin K in the diet.

As with most of the fat-soluble vitamins, losses during cooking and processing are very low.

Review

1.   **Vitamins** – organic substances required in very small amounts

2.   **Water-soluble vitamins:**
     Ascorbic acid (C)
     'B' complex

3.   **Fat-soluble vitamins:**
     Retinol (A)
     Calciferols (D)
     Tocopherols (E)
     Naphthoquinones (K)

4. Ascorbic acid (C)

- deficiency causes 'scurvy'
- destroyed by heat, leached by water, oxidised readily, stable in acid
- natural reducing agent
- considerable losses in cooking and processing

5. Thiamin (B_1)

- deficiency causes 'beri-beri'
- destroyed by heat, leached by water, sulphur compounds cause breakdown
- lost in cooking and processing, particularly in milling cereals, eg rice

6. Riboflavin (B_2)

- deficiency causes skin problems – but rare
- heat-stable but destroyed in alkali
- useful yellow colour

7. Pyridoxine

- deficiency rare
- lost in milling cereals
- destroyed by heat, particularly in alkali
- susceptible to light

8. Cyanocobalamin (B_{12})

- deficiency causes pernicious anaemia
- found in animal products only
- heat-stable, not lost in cooking or processing

9. Nicotinic acid (niacin)

- deficiency causes pellagra
- soluble in water but usually unaffected by cooking or processing
- often unavailable in a food, eg in cereals

10. Pantothenic acid

- deficiency rare
- destroyed by dry heat, acids and alkalis

11. Biotin

- deficiency rare
- unaffected by cooking and processing
- bound by egg protein – avidin

12. Folic acid

- deficiency causes anaemia
- may be deficient in some UK diets
- destroyed by heat processing, particularly in presence of air

13. Retinol (vitamin A)
– deficiency causes 'night blindness'
– excessive amounts cause poisoning
– carotenoids converted to retinol but require β-*ionone* ring in their structure
– stable in cooking and processing, but oxidised in dried products easily

14. Calciferols (vitamin D)
– deficiency causes 'rickets'
– synthesised by action of sunlight on skin or from fish liver oils
– related to cholesterol
– unaffected by cooking and processing
– excessive intake causes poisoning

15. Tocopherols (vitamin E)
– deficiency rare in human
– sterility in rats
– natural antioxidant
– extravagant health claims
– lost in processing by oxidation

16. Naphthoquinones (vitamin K)
– deficiency – uncontrolled bleeding
– one form produced in gut by bacteria
– little loss in cooking or processing

Practical exercise: *Vitamins*

(It is only possible to carry out simple experiments on ascorbic acid, which is readily available.)

Determination of Vitamin C

The dye 2 : 6 dichlorophenol indophenol (DCPIP) is discolourized by ascorbic acid (and other reducing agents). The dye is available in special tablets, and usually one tablet, when made into solution, is decolourized by 1 mg of ascorbic acid.

Ascorbic acid can be determined in fruit juices, natural fruits and other fruit products.

Use a pipette and put 10 cm^3 of juice into a standard 100 cm^3 flask; add about 50 cm^3 of acetic acid and make up to volume with distilled water. Make up a dye solution as per label instructions and fill into a burette.

Pipette 10 cm^3 of prepared juice into a flask and add a few drops of acetone (this is to remove sulphur dioxide which interferes with the titration). Titrate the dye, which is decolourized, until all the ascorbic acid in the juice is used up, and the end-point is when a feint pink colour persists. Calculate the ascorbic acid content of the juice.

1.7 Pigments

The attractive and often unique colours of a particular food are due to the presence of small amounts of pigment molecules. The colour plays an important role in the appeal of a food and is of equal consequence in manufactured foods, such as convenience products, as in natural foods. Colour plays a part, with other factors, in the overall organoleptic assessment of the food. If, for example, a strawberry flavoured ice-cream is coloured green it will confuse many tasters. Colour is a good indicator of the freshness of foods as microbial deterioration or other spoilage is always accompanied by a change in colour. Fruit and vegetables indicate their ripeness by a change in colour.

Food colours are usually divided into those which occur naturally and those which are synthetic. However, this is a little confusing as some natural food colours can be readily synthesised and the natural and synthetic colours are indistinguishable. Synthetic or manufactured colours grew out of the dye industry, initially developed for textiles. Colours were selected for foods because they were low in perhaps lead or arsenic. Legislation has subsequently demanded very high levels of purity and high colour potential. Much criticism is levelled at these colours by those opposed to additives. However, the main synthetic food colours have been subjected to extensive toxicological testing and are only approved when known to be safe. These colours offer advantages to the food manufacturer which include: lower price, consistency and strength of colour, known performance during food processing, and, usually, ready availability.

Let us, therefore, cease to use this word 'synthetic' and split the colours into those naturally present in foods, and those added to foods.

1.7.1 Colours naturally present in foods

Colour compounds are often complex structures. Pigments naturally occurring in foods are often blends of a number of different pigments and a food sometimes has different pigments at different stages of its existence in the 'unharvested state'. For example, a tomato is coloured green with chlorophyll when unripe; this green colour, and chlorophyll, gradually disappears on ripening whilst the red carotenoid, lycopene, is synthesised. As animals age their fat becomes more yellow as carotenoids become dissolved in the fat over a long period of time.

In Table 1.11 the naturally occurring pigments are given, with the main examples of each; and where they can be found. (Please remember that these compounds can be extracted or synthesised in some cases and added to the foodstuffs, but it is convenient to classify them as naturally occurring.)

Table 1.11 Naturally occurring colours

Main group	Examples	Colours	Occurrence
Isoprenoid derivatives	Carotenoids: carotenes	orange	carrots, apricot, fish oil
	lycopene	red	tomatoes, water melon
	xanthophylls	yellow	citrus juice
Tetrapyrrole derivatives	Chlorophylls	green	vegetables, unripe fruit
	Pheophytin	grey/brown	produced when chlorophyll is heated
	Haemoglobin	dull red	blood (no oxygen)
	Oxyhaemoglobin	bright red	blood (with oxygen)
	Myoglobin	purple/red	muscle/meat
	Oxymyoglobin	bright red	muscle (with oxygen)
Benzopyran derivatives	Anthocyanins	red/blue/ purple	blackcurrants, flowers, cherries
	Flavones	white/ yellow	potato, apples
	Tannins	brown	tea

1.7.1.1 Isoprenoid derivatives – Carotenoids

These compounds are derived from isoprene (C_5H_8) and all true carotenoids have 40 carbon atoms, ie eight isoprene units. These pigments vary from yellow to red and are mainly fat-soluble. The pigments are structurally related to retinol, vitamin A.

Carotenoids must have a β-ionone ring to have pro-vitamin A activity, (see Section 1.6.2.1). In Figures 1.52 and 1.53 are the structures of the most common carotenoid, β-carotene, showing two β-ionone rings, and lycopene showing two open rings, with no pro-vitamin A activity. β-carotene is always accompanied by other carotenoids, for example in carrots by α- and γ-carotene.

The carotenoids are generally stable to heat during processing and cooking but can be readily oxidised in dehydrated foods to form colourless compounds. The alternating double-single bonds, conjugated double bond system, explains the intense colour of these compounds and their stability in most situations.

Carotenoids can readily be seen as the dominant colours of tomatoes,

Figure 1.52 Structure of β-carotene

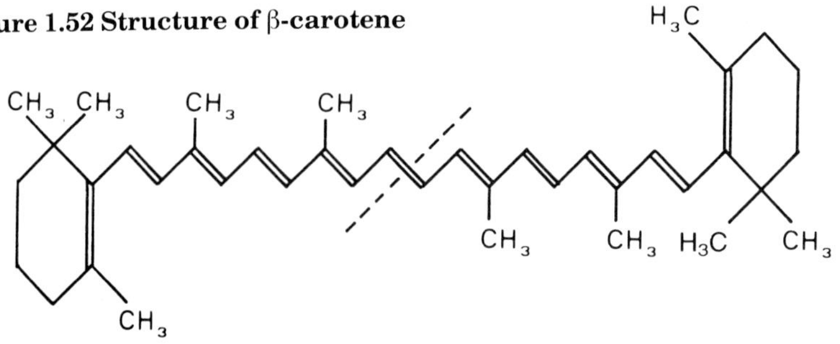

Figure 1.53 Structure of lycopene

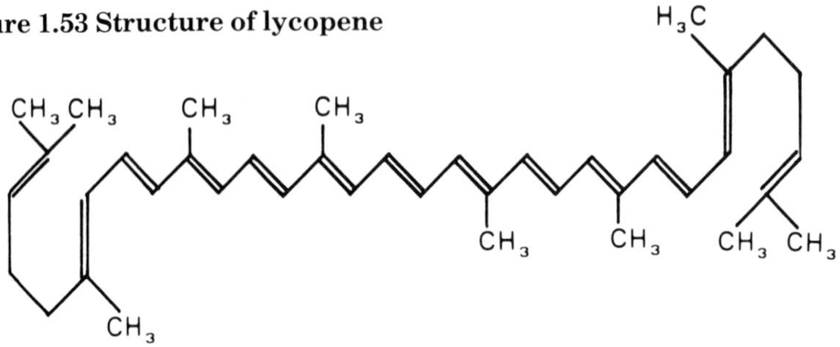

water melons, peaches, peppers, carrots, apricots and spices such as saffron. However, one cannot assume that a yellow, orange or red colour is a carotenoid as a number of the more reddy pigments can easily look like some anthocyanins. The carotenoids are often masked by other pigments, particularly by chlorophyll. When a fruit ripens chlorophyll is broken down and the yellow/red carotenoid is exposed to give the ripe appearance of the fruit. Lycopene is an exception as it is synthesised by tomatoes as chlorophyll is broken down during ripening.

Considerable quantities of carotenoids are extracted and used as colours for 'manufactured' foods. However, as the carotenoids are insoluble in water they are often converted into emulsions in water before mixing with a food. The most important applications of carotenoids are in colouring soft drinks, jellies, boiled sweets, desserts and yogurts. Carotene as a suspension in oil can be used to colour margarine, with the advantage that it adds pro-vitamin A activity to the product.

Carotenoids which contain hydroxyl groups (−OH) are called *xanthophylls*. They are often mixed with carotenes in food pigments and tend to have a light yellow colour. A typical example is cryptoxanthin, which, in effect, is β-carotene with one hydroxyl group (OH) attached to the second β-ionone ring on carbon atom 3. Cryptoxanthin is the chief pigment of maize, paprika and mandarin orange.

In Table 1.12 a summary is given of some of the combinations of carotenoids to give the colour of a particular food product.

Table 1.12 Combinations of carotenoids in foods

Food	Carotenoids
Orange	β-carotene, lycopene, cryptoxanthin, xanthophyll, violaxanthin
Red pepper (Chilli)	α-carotene, β-carotene, capxanthin
Carrot	α-carotene, β-carotene, γ-carotene, xanthophyll
Maize	α, β, γ-carotenes, zeaxanthin, crytoxanthin, xanthophyll and several others
Peach	β-carotene, crytoxanthin, xanthophyll, zeaxanthin

1.7.1.2 Tetrapyrrole derivatives

Chlorophylls

Chlorophylls are the most common pigments found in foods as they occur in leaves, unripe fruit and many vegetables. Chlorophylls occur within the cells in small bodies, the chloroplasts, and usually are combined with proteins. There are two chlorophylls, a and b, and usually there is three times as much a as b. Chlorophylls are large molecules composed of four pyrrole rings (hence tetrapyrrole) held together by methene bridges ($-CH=$). In the simplified structure of a chlorophyll given in Figure 1.54, the four pyrrole rings can be seen and in the centre an atom of magnesium is held. Around the outside of the structure in positions 1 to 8 there are different groups such as methyl, ethyl, vinyl, and a specific alcohol is attached at position 7 called phytol.

Figure 1.54 Structure of chlorophyll (simplified)

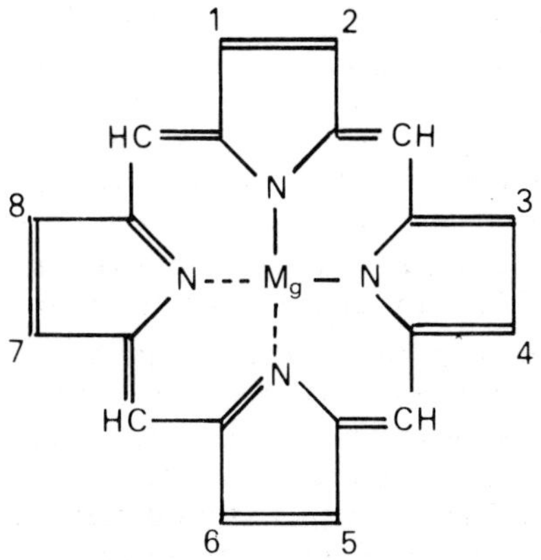

Chlorophyll b differs from a in that it has an aldehyde group (CHO) in position 3 instead of a methyl group (CH_3).

The chlorophylls are very unstable molecules when the living plant cell is killed during cooking or processing. Thus the green colour of chlorophylls is readily lost during heat processing. As the chlorophylls exist as protein complexes in the living cell, when the cell is killed by cooking or processing the protein is denatured and the chlorophyll released. Acids present in the food, or produced by heat, are able to substitute two atoms of hydrogen for the magnesium in the centre of the chlorophyll. The chlorophyll changes to an olive green colour or even to brown as the substitution of magnesium by hydrogen produces a substance called pheophytin. Pheophytin has the same tetrapyrrole structure as chlorophyll but contains hydrogen and not magnesium in the centre of the structure.

Obviously, in the absence of acid this reaction will not occur, or only slightly. Sodium hydrogen carbonate (bicarbonate of soda) can be added to vegetables to make them slightly alkaline and thus reduce chlorophyll decomposition, keeping the bright green colour. However, as pointed out in Section 1.6, a number of vitamins, particularly C and B_1, are easily destroyed under alkaline conditions.

A number of vegetables such as peas, spinach, sprouts and cabbage, produce a number of acids during heat processing, such that the pH may fall from about 6·6 to 6·1. These acids will accelerate the decomposition of chlorophyll unless the acidity is decreased by the addition of bicarbonate.

Some metals react with chlorophylls to form bright green compounds. Iron III, zinc and copper II ions will replace the magnesium in chlorophyll to produce stable green products. Unscrupulous manufacturers in the past have added copper salts to canned vegetables to improve their colour. Obviously this practice could lead to severe poisoning.

Haemoglobin and myoglobin

It is a remarkable coincidence of nature that the red pigment of many animals, haemoglobin, is structurally very similar to the green pigment, chlorophyll, of plants. Haemoglobin, however, contains a tetrapyrrole derivative, called haem, with iron II and not Mg in the centre of the structure. In blood, haemoglobin exists as four haem units, joined to one molecule of protein.

The groups attached at positions 1 to 8 are methyl, vinyl, methyl, vinyl, methyl, propanoic acid, propanoic acid and methyl, in that order.

Myoglobin is the principal pigment present in muscle, and it is composed of only one haem unit and one protein molecule.

The unique property of haemoglobin and myoglobin is the capacity to bind a molecule of oxygen but without oxidising the iron II (ferrous) to iron III (ferric) in the centre of the structure. The reaction is readily reversible and is the life-giving system which transfers oxygen from the

Figure 1.55 Structure of haem

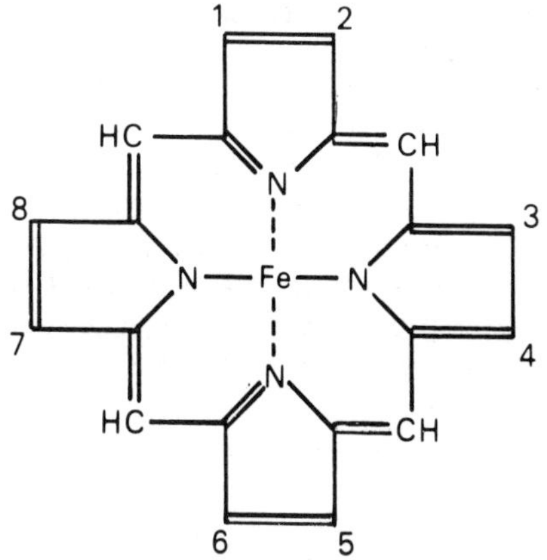

lungs to the tissues. Haemoglobin and myoglobin are a dull purple/red colour but change to a bright red colour when combined with oxygen to form oxyhaemoglobin and oxymyoglobin respectively. This can be seen readily in a joint of beef, the side exposed to the air is bright red but the side next to the plate or packing tray is starved of oxygen and therefore dull red.

If the haem units become detached from the proteins in either myoglobin or haemoglobin they are susceptible to irreversible changes. In this situation, perhaps caused by heat, acids or oxidising agents, the iron is converted from iron II to iron III and the pigment becomes a brown colour due to the formation of methaemoglobin or metmyoglobin. This is readily shown when cooking meat. Heat denatures the protein and allows the haem to disengage itself from the protein, and thus the iron the haem contains is readily oxidised to the iron III form. Sometimes slight changes occur in the haem structure which result in the formation of green pigments. This can sometimes be seen at the edge of slices of cooked meat.

In cured meat a pink colour is produced as nitrosomyoglobin is produced in the product by nitrogen II oxide (nitric oxide (NO)) replacing oxygen held by iron in the structure. The nitrogen II oxide is produced from nitrates and nitrites used in the curing operation.

1.7.1.3 Benzopyran derivatives

Anthocyanins

These water-soluble pigments are responsible for many of the bright colours of flowers and many fruits. Unlike many other pigments, the anthocyanins change their colours under various conditions, particularly changing pH. Colours range from red to blue and purple.

Anthocyanins are made up of a complex ring structure (see Figure 1.56) and attached to this is a sugar or a number of sugars. The ring structure is called an *anthocyanidin*.

Figure 1.56 Structure of an anthocyanidin

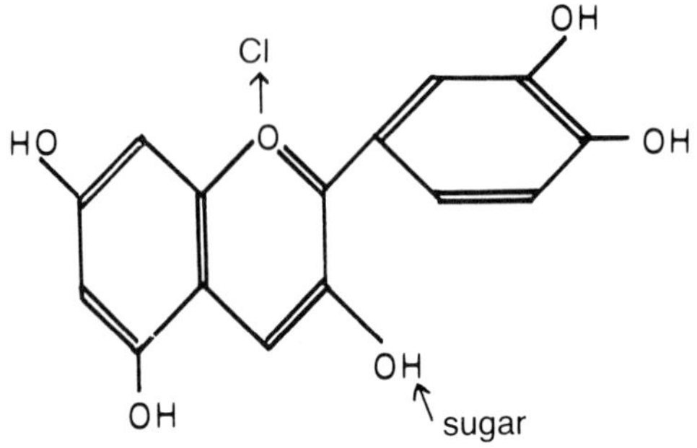

An anthocyanin = anthocyanidin + sugar(s).

If the number of hydroxyl groups (OH) is greater in the anthocyanidin the resulting anthocyanin is more blue in colour. If the number of methoxyl groups (OCH_3) is increased the colour becomes more red. The anthocyanidins have interesting names based on flowers, for example, peonidin produces a red anthocyanin and delphinidin a blue one.

The anthocyanins are good pH indicators as they are red at low pH (acid) and blue at high pH (alkaline). The colour can be lost if sulphur dioxide or sulphites are used to preserve a product, for example fruit pulp, as the pigments are decolourised.

The pigments can combine with metals, particularly iron, tin and aluminium. Some fruits canned in unlacquered cans produce a greyish sludge of anthocyanin which has combined with the metal of the can.

Anthocyanins are very stable in acid conditions and therefore can be extracted and used in acid foods to achieve strong red colours. Canned strawberries can have citric acid added to lower the pH to almost 3 so that an excellent red colour is obtained. Anthocyanins can be added to soft drinks, confectionery, jellies, yogurts and desserts.

There are a number of *flavone* derivatives which have a similar structure to anthocyanidins but are usually colourless, yellow or orange pigments.

(*NB* most bright yellow and orange fruits are coloured by carotenoids.)

An example of this type of pigment is quercetin which is found in onion skins, hops and tea.

The tannins are another similar group of pigments which are complicated mixtures giving a red or brown colour. The significance of tannins in a number of foods is due to the astringency which can readily be appreciated when drinking strong tea. Tannins give body, fullness of flavour and help preserve some red wines, particularly wines from the Bordeaux region, such as clarets, famed for their longevity.

1.7.2 Colours as food additives

Food dyes (or synthetic colours) are used throughout the world, although the USA and Europe are the main centres for toxicological tests. In the EC there is a special Food Colourants Working Party which submits detailed suggestions on behalf of food manufacturers and a list of permitted colours is published and reviewed by the EC at regular intervals. Many of the colours have been known for many years and the colour industry is an established one aware of consumer reactions and market requirements. Most of the compounds are complex nitrogen-containing structures derived from coal-tar. It cannot be denied that a colourful product is attractive and so there will always be a demand for such colours, particularly as natural colours may be limited, expensive or unavailable.

Table 1.13 (page 85) lists the currently approved colours in the EC, the colours are split into naturally occurring and those of dye origin.

Tartrazine has been implicated in an allergic reaction of some people to coloured soft drinks, such as orange squashes. The allergy shows itself in coughing and itching about the neck. It has also been implicated as a cause of hyperactivity in children.

Many colours have been withdrawn from use over the last few years and it can be assumed that any suspected or implicated in any toxicological disease will be withdrawn.

Review

1. **Colour** good indicator of the freshness of food

2. **Pigments naturally occurring**

Isoprenoid   derivatives
Tetrapyrrole   ,,
Benzopyran   ,,

3. **Isoprenoid derivatives**

   – carotenoids – contain 40 carbon atoms
   – yellow/orange/red
   – require complete β-ionone ring to have pro-vitamin A activity
   – most common β-carotene
   – stable to heat but can be oxidised
   – if contain hydroxyl group (OH) known as *xanthophylls*

4. Tetrapyrrole derivatives

– chlorophylls
– 4 pyrrole rings held together by methene (–CH=) bridges
– atom of magnesium held in centre of structure
– very unstable, readily destroyed in cooking
– susceptible to acids but stable in alkalis
– pheophytin, grey/brown formed when magnesium replaced by hydrogen due to acids in cooking

5. Haemoglobin

– red pigment of blood
– similarly myoglobin – red pigment of muscle
– similar structure to chlorophyll but different side chains and iron II in centre of structure
– oxyhaemoglobin and oxymyoglobin bright red, if lose oxygen become dull red or purple haemoglobin or myoglobin
– pink colour in cured meats due to nitrosomyoglobin

6. Benzopyran derivatives

– anthocyanins
– red to purple water-soluble pigments found in flowers and many fruit
– anthocyanin = anthocyanidin + sugar(s)
– anthocyanidins, named after flowers, eg peonidin – deep red
– anthocyanins red in acid, blue in alkali
– react with metals to form grey compounds

7. Colours as food additives

– most derived from coal tar
– permitted list only may be used in foods
– very stable, known colour intensity and cheaper than natural colours

Practical exercises: *Pigments*

Use a range of fruits and vegetables to test the following pigments.

Note colour changes produced.

1. Chlorophylls

Boil samples of green vegetables in the following:
  (a)   distilled water
  (b)   dilute alkali
  (c)   dilute acid
  (d)   dilute copper sulphate solution

2. Carotenoids (eg in carrots)

  (a)   Dice the carrots and boil for 10, 20 and 30 minutes.

  (b)   Pressure cook samples at various pressures and for various times.

  (c)   Compare colours of diced carrots after each process.

3. **Anthocyanins** (eg in red fruits: strawberry, plum, raspberry)

Repeat (a), (b) and (c) above. (Question 1).

    (d)   add a few drops of a 5% stannous chloride solution

    (e)   add a few drops of a 5% ferric (iron III) chloride solution.

Table 1.13 Permitted food colours

Colour	Natural/Dye	EC Code	Name
YELLOW	Natural	E.100 E.101	Circumin (from turmeric) Riboflavin (vitamin B_2)
	Dye	E.102 E.104 E.110	Tartrazine Quinoline yellow Sunset yellow
RED	Natural	E.120 E.162 E.172	Cochineal Beetroot red Iron oxide
	Dye	E.122 E.123 E.124 E.127	Carmoisine Amaranth Ponceau 4R Erythrosine BS
BLUE	Dye	E.131 E.132	Patent blue V Indigo carmine
GREEN	Natural	E.140 E.141	Chlorophylls Copper chlorophylls
	Dye	E.142	Green S
BROWN	Natural	E.150	Caramel
BLACK	Natural	E.153	Carbon black
	Dye	E.151	Brilliant black
WHITE	—	E.171	Titanium oxide
RED-BLUE	Natural	E.163	Anthocyanins
YELLOW/ ORANGE/ RED	Natural	E.160 E.161	Carotenoids Xanthophylls
OTHERS	—	E.173 E.174 E.175 E.180	Aluminium Silver Gold Pigment rubine

(*NB* Natural is used here to distinguish from a dye, but some natural products could be manufactured.)

1.8 Flavours

The flavour of foods is one of the delights of eating. Food without a certain flavour level is usually considered dull and unappetising and restaurants producing such dishes quickly lose their clientele. Flavour is a combination of taste and smell and in many cases also of 'mouthfeel'.

1.8.1 Taste

There are only four true tastes and other 'tastes' or flavours are in fact odours. The four tastes – salt, sweet, sour and bitter – are detected by taste buds on the tongue, the pharynx and soft palate. The taste can be detected if the particular substance is in solution in the saliva or in the natural juices of the food. A sweet taste is detected on the tip of the tongue; sour is detected on the sides; salt on the sides and tip; and bitter at the back of the tongue and on the pharynx.

The *salt taste* is a property of the electrolytes, ie low molecular weight ionised salts, particularly the halides such as sodium chloride. The order of saltiness is as follows: chlorides, bromides, iodides, sulphates and nitrates. Sodium salts are obviously salty whereas potassium salts can be bitter and unpleasant. This is probably why salt-substitutes, ie sodium substitutes, for people on a salt-free diet, are not considered to be particularly salty and take some time to be accepted.

A *sweet taste* is, of course, typical of the low-molecular carbohydrates such as fructose, glucose and sucrose. Sucrose is an ideal sweetener, with no after-taste and it can add body to soft drinks. However, particularly in confectionery, it is not sweet enough, and it will cause tooth decay, weight increase and possible heart problems. Many other substances show a degree of sweetness and some are many times as sweet as sucrose. Saccharin, up to 500 times as sweet, has been used for many years as a sucrose substitute or in sweeter products. Unfortunately it can have a bitter after-taste and is not particularly stable in heat processed foods.

Figure 1.57 Structure of saccharin

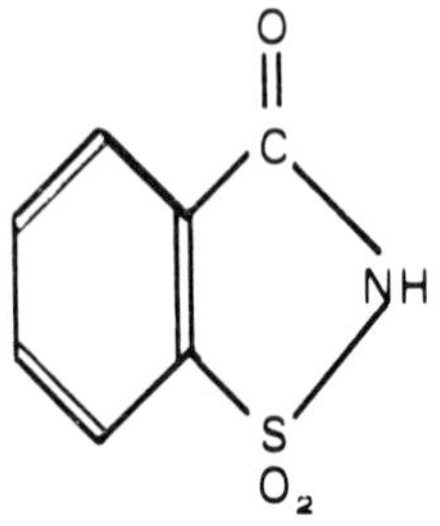

Saccharin, although not proven, has been implicated in some cases of bladder cancer. However, this was discovered using test animals, and in a similar manner the sweetener *cyclamate* was banned as it produced cancer in rats.

Some combinations of amino acids are very sweet. The new sweetener, *aspartame,* is made from phenylalanine and aspartic acid. It is nearly 200 times as sweet as sucrose, and is now approved for general use.

Sourness is the taste of acid and is a property of the hydrogen ion (H^+). Food acids are organic acids and are relatively weak or sometimes undissociated (unionized) acids. Some foods require an acid taste to be acceptable, particularly citrus fruits, and to a lesser extent apples and fermented products such as yoghurt.

Bitterness is a property of a number of organic and inorganic compounds, particularly the alkaloids and on occasions substances containing magnesium, calcium and ammonium ions. Some bitter alkaloids include quinine, in tonic water, and caffeine in coffee.

1.8.2 Odour

Odours are detected in the nose when compounds in minute quantities come in contact with the olfactory nerve endings. Thousands of different compounds can be detected and they are responsible for giving the characteristic and often subtle 'flavour' of many foods. Food flavour, therefore, is in reality a combination of many odoriferous compounds. The essential requirement for such compounds is that they should be volatile. When a food is eaten the odoriferous compounds change to gases and diffuse through the pharynx up into the nose. We can, of course, smell an odour prior to eating a food and the memory of this odour, which is always good in most people, stimulates the olfactory senses with a resulting production of saliva, ie the mouth 'waters'.

Food flavours are usually complex mixtures of hydrocarbons, alcohols, acids, aldehydes, ketones and esters. The most common types of food flavours, found in fruits and vegetables, are called *essential oils*. This term is somewhat confusing as it is derived from 'essence' and does not mean these oils are essential nutrients for the body. Essential oils occur in most parts of plants and can be readily extracted by pressure or steam distillation. Many of the oils are found in special oil sacs, for example, the skin or oranges contain essential oil which sprays from these sacs as the orange is peeled. An enormous industry has developed producing oils, particularly in small island states of the West Indies and Indian Ocean, where they constitute the entire source of foreign exchange. Examples of essential oils include oil of almond, clove, garlic, ginger, lemon, lime, mace, orange and thyme. Less volatile flavour substances can be extracted from plants, particularly spices and herbs, by the use of solvents such as acetone and propan-2-ol. These flavour extracts are called *oleoresins*.

The chemical constituents of essential oils are either terpenoids or other compounds.

1.8.2.1 Terpenoids

The terpenoids are a very large group of substances which, like the carotenoids, are based on isoprene (C_5H_8). The monoterpenes have two isoprene units (hence their formula of ($C_{10}H_{16}$)); the sesquiterpenes have three ($C_{15}H_{24}$); and the diterpenes have four units ($C_{20}H_{32}$).

The monoterpenes are the most common group, having the strongest odours. The odour of terpenes is increased if oxygen is included in their structure. As the number of carbon atoms increases, volatility and odour tend to decrease. The monoterpenes can be divided into three main groups: the acyclic, monocyclic and bicyclic monoterpenes.

Acyclic monoterpenes do not possess a ring structure, but readily close to form a monocyclic monoterpene in certain conditions particularly when heated in the presence of acid, for example in fruit juice. Most of the acyclic monoterpenes have pleasant odours and an oxygenated derivative, citral (actually two similar compounds) gives the flavour to lemon oil.

Figure 1.58 Acyclic monoterpenes

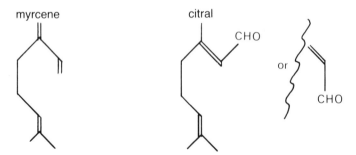

Monocyclic monoterpenes represent the most stable form of monoterpenes, and often during storage, processing and extraction of flavours from fruits other monoterpenes are converted to this group. Lime oil clearly shows the results of such changes as this. Expressed lime oil from the skin of the fruit is similar to lemon oil in odour, but when lime oil is steam-distilled from the fruit pulp and skin, a method commonly used to increase yield, the monoterpenes undergo many changes, for example citral is partially converted into a monocyclic monoterpene, limonene. Limonene is a common example of this group and is found in many essential oils. A common alcohol, which can be derived from limonene, is α-terpineol, again found in lime oil particularly when it is badly stored or old.

Figure 1.59 Monocyclic monoterpenes

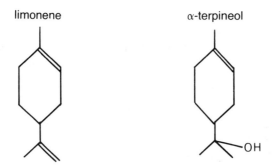

limonene   α-terpineol

Bicyclic monoterpenes constitute a large group of monoterpenes having an unusual structure of one four-sided ring within a six-sided ring. The pinenes have the smell of pine disinfectant but occur in small amounts in many essential oils. There are in this group a number of oxygenated derivatives particularly alcohols.

Figure 1.60 Bicyclic monoterpenes

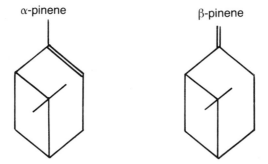

α-pinene   β-pinene

1.8.2.2 Other odour compounds

No other single group of compounds in odours of foods is as large as the terpenoids. However, a wide range of substances can be found together with the terpenoids in essential oils and these include hydrocarbons, alcohols, aldehydes, esters, ketones, acids and ethers.

Some flavours, although often very complex mixtures of compounds, depend on one particular substance to give that characteristic flavour. Synthetic flavours used to be made from this particular substance and so were generally poor in comparison with the natural product. Esters are particularly common in fruit flavours and some of the common ones are listed in Table 1.14.

Table 1.14 Principal flavouring agents in foods

Food	Flavouring agent	Chemical nature
Almond	benzaldehyde	aldehyde
Banana	amyl acetate	ester
Butter	diacetyl	ketone
Cloves	eugenol	alcohol
Coconut	an aldehyde	aldehyde of 14 C. atoms
Grape	methyl anthranilate	ester
Lemon	citral	aldehyde
Mint	menthol	alcohol
Pear	ethyl acetate	ester
Pineapple	allyl caproate	ester
Raspberry	ethyl formate	ester

It must be pointed out that to obtain the full characteristic flavour many other substances are needed besides the ones listed in the table and these substances also occur themselves in many other flavours.

Vegetables have a limited range of volatile odorous compounds but a number of flavours are developed in cooking. The onion family is rich in sulphur containing compounds, such as alliin, which is converted by enzymes to diallyl thiosulphinate which has the typical strong smell. The cabbage family, similarly, is rich in sulphur-containing compounds which are released on cooking.

1.8.3 Mouthfeel

Mouthfeel is often used to indicate the texture of a food when eaten, but it can also be used to indicate sensations such as 'tingling, hot and watery'. The olfactory nerve endings initiate the sensation of smell and the detection of odour, but some nerves in the skin, tongue and cheeks are sensitive to other sensations. Smell ammonia and a 'tingling', quite irritating sensation will be experienced. Many spices and peppers, such as chilli, are described as hot, whereas peppermint may be described as cool.

The texture of food will influence the flavour. Smooth products taste differently from rougher similar products, for example, the sweetness of confectionery may be affected by sucrose crystal size.

1.8.4 Food flavours

Naturally occurring flavours are often extracted from fruits, spices and herbs by the use of solvents or distillation and then used in manufactured food products. Citrus oils can easily be expressed from the peel of the fruit and special machines have been developed for this. A traditional tool in the West Indies, called an écuelle, is used for expressing citrus oils. It is basically a copper funnel to which are attached a number of studs, on these the fruit is rubbed. Pieces of peel, some juice and essential oil run down the funnel and are collected. After standing, the oil comes to the surface of the mixture and is removed. Steam distillation can be used to increase yield, but as mentioned earlier, in limes for example, flavour changes will result.

Figure 1.61 An écuelle

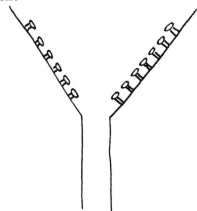

Naturally occurring flavours have similar disadvantages to natural colours in being variable, sometimes expensive and often unstable.

Synthetic flavours often meet consumer resistance but have definite advantages for the food processor. Such flavours have been used for many years and are being blended on to sugar, salt, flour or glucose as a means of adding them easily to a dried food product or dissolved in alcohol, eg ethanol or propanol, for use in liquid foods. Originally, just esters were used to flavour food products but better analytical techniques have helped to discover a wide range of flavour compounds.

Gas chromatography has probably made the greatest contribution to the study of flavours in being able to separate all the minute components of a flavour which can subsequently be identified by other techniques such as mass spectrometry. A flavour manufacturer cannot analyse a natural flavour then mix all the components, perhaps 100 in all, from various sources to produce an artificial flavour. Normally the main flavouring agents are recognised and these, perhaps with some natural extracts, are blended to make the flavour. In a highly competitive industry the discovery of a particular flavour component is a closely guarded secret on occasions.

Flavours alter in processing and in cooking. A good flavour producer will recognise this and build into his flavour a capacity to change to the desired flavour after processing. A flavour, therefore, might not seem to be correct, but after baking, for example, it may have exactly the right flavour notes. Considerable research is undertaken on flavour precursors for this reason. Intermediate products from the Maillard reaction have, for example, a tendency to develop a 'meaty' flavour.

1.8.5 Flavour modifiers

There are a number of substances used in foods which are capable of enhancing and sometimes reducing the level of a flavour. Salt is a flavour enhancer for some foods and soy sauce has been used for many years in China and Japan. It was found that soy sauce was rich in *monosodium glutamate* (MSG), a derivative of the amino acid glutamic acid, Figure 1.62.

Figure 1.62 Monosodium glutamate

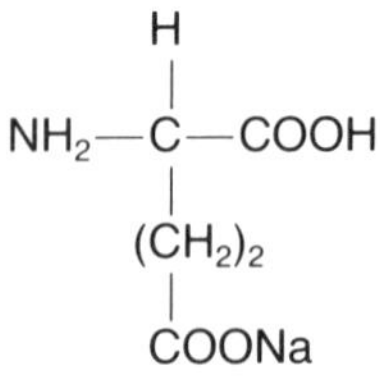

$$NH_2-\overset{\displaystyle H}{\underset{\displaystyle (CH_2)_2}{C}}-COOH$$
$$COONa$$

Large quantities of MSG have been used for many years in the food industry and in cooking. Excessive use of MSG can cause dizziness and sickness, first recognised in Chinese cooking and called the 'Chinese Restaurant syndrome'. Food, to which has been added MSG, is often preferred by taste panellists to untreated food. MSG has a 'warm' slightly sweet flavour and it noticeably enhances meat flavours and some vegetable flavours, but generally has no effect on or slightly suppresses sweet flavours. It tends to 'round-off' flavours and will suppress certain strong flavours, such as that of onion. MSG is used in soups, both canned and dried, tinned meats and vegetables.

Ribonucleotides, produced by micro-organisms, have a similar effect to MSG but can be used in very small quantities. Ribonucleotides are now used often as blends with MSG and are formed from ribose, phosphoric acid and a base such as inosine or guanine. They are, of course, naturally occurring particularly in yeast and meat.

1.8.6 Taste panels

No instruments are available which can measure the sensation of flavour. Although gas chromatographs can separate flavour components, which then can be individually smelt, the overall flavour cannot be measured. Some food industries employ expert tasters with many years of experience, particularly for wine, tea, whisky and some spices. The average food factory has to employ a taste panel to evaluate its products and compare their flavour with those of a competitor. There are a number of standard tasting techniques which can give reliable results if conducted in a properly controlled environment where tasters are not influenced by outside factors or by each other.

Difference tests are often used with either a pair of unknown food products, or three in a triangle, in which two are identical and the third is to be separated. In *ranking* a series of samples have to be put in order by the panel according to increasing sweetness, saltiness, flavour or any other characteristic.

A *flavour profile* is an elaborate evaluation of a flavour by a trained panel. The flavour is broken down into its 'components' which are described by any term which accurately describes it, such as mushroom-like, catty, horsey, eggy and so on. The individual flavour notes are considered; the order of their appearance in the mouth; their strength; and the presence of any after-taste.

Taste panels have many uses but are subject to error and must contain a representative segment of the population to obtain significant results.

Review

1. Taste – 4 tastes – salt, sweet, sour, bitter

2. Sweet – detected on tip of tongue
   Sour – detected on sides of tongue
   Salt – detected on sides and tip of tongue
   Bitter – detected on back of tongue

3. Saltiness property of electrolytes, eg sodium chloride

4. Salt-substitutes – potassium based, eg potassium chloride

5. Sweeteners – saccharin – bitter after taste
   – up to 500 times as sweet as sucrose
   – aspartame – new sweetener, 200 times as sweet as sucrose
   – made from phenylalanine and aspartic acid

6. Sourness – due to acid
   – essential for some products, eg citrus fruits

7. Bitterness – due to alkaloids, eg quinine

8. Odour – large number of compounds give characteristic flavour of foods

9. Essential oils – blends of many flavour compounds, naturally occurring

10. Oleoresins – extracted by solvents from spices and other products

11. Terpenoids – large group of flavour compounds
    – monoterpenes – because of volatility, are most common flavours

12. Monoterpenes – Acyclic, eg citral
    – Monocyclic, eg limonene
    – Bicyclic, eg pinene

13. Some non-terpenoid flavour compounds – esters, eg amyl acetate in bananas, diacetyl in butter

14. Vegetables – limited range of odourous compounds
    – sulphur compounds released on cooking

15. Flavour modifiers – monosodium glutamate (MSG) and ribonucleotides
    – enhance meat and vegetable flavours, suppress harsh flavours eg onion

16. Taste panels – comparison and difference tests:
    eg triangle test
        ranking
        flavour profile

Practical exercises: *Flavours*

1. Citrus fruits

1. Examine the peel of a citrus fruit with a hand-lens and under a low-power microscope. Identify the oil sacs containing the essential oil.

2. Extract oil from the oil sacs of a citrus fruit by rubbing the fruit on a grater. Allow to settle and separate the small amount of oil from the skin debris.

3. Crush a fresh lime and note the smell of the essential oil. Gently heat the crushed fruit and note the change in aroma from lemon-like to the typical lime aroma.

2. Taste panels

Set up a taste panel and prepare samples to be tested by difference (2 samples) or by triangle testing (3 samples, 2 identical).

Vary one ingredient in a recipe, for example, salt, sugar, MSG and flavouring. Ask panelists to pick out the one they prefer, or which one is different.

The panel, when experienced, may be able to undertake a 'flavour profile'. Manufactured products may be compared for individual flavour notes; order of their appearance in the mouth; their strength; and the presence of after-tastes. Instant soups are a good starting point.

1.9 Additives

In support of additives

Additives have allowed the development of a wide range of interesting and often 'convenient' food products. The food industry must provide food that is appetising, nutritious and visually attractive. In many cases without the use of additives food quickly deteriorates. Over 5000 substances have been used in the food industry as additives and about half of these are flavour compounds contributing to a total market of £150 million per annum. When approved within the EC an additive is given an 'E number'. Any additives with just a number and no letter 'E' are approved within the UK only.

In Table 1.15 are listed the various groups of additives and their possible uses in foods, some of which will be obvious.

Table 1.15 Additives

Additive group	Use
Acids	Flavour, sometimes preservation
Anticaking agents	To prevent lumps forming in powders
Antioxidants	Prevent fat rancidity (see section 1.3.4)
Buffers	Control pH
Colours	To attract the consumer (see section 1.7)
Emulsifiers	Emulsification of oil in water (see section 1.3.1.5)
Enzymes	Many uses – eg invert sugar production (see section 1.4.2.2)
Firming and crisping agents	Control texture
Flavours	To improve a natural product or flavour a manufactured food (see section 1.8)
Humectants	To aid moisture retention
Nutritive additives	To improve vitamin, mineral or protein content of a food
Preservatives	To prevent microbial spoilage in some products
Sequestrants	Bind up metals to prevent their reaction with other food components

Acids

The majority of foods are acidic or neutral and very few are alkaline, eg egg white. A food produced from a number of different food components by a manufacturer must have some acid added to help give the right taste and degree of sourness. Instant desserts, soups, jellies, sauces, soft drinks and confectionery normally require some acid to be added to

Figure 1.63 Citric acid

CH₂COOH
|
C(OH)COOH
|
CH₂COOH

them. Citric acid is a common acid used in food and which occurs naturally, particularly in the citrus fruits.

In addition to its acidity, citric acid, when ionized, has the ability to act as a sequestrant or chelating agent and bind up metal ions, such as calcium (see Figure 1.67). This property can be of use in many products, particularly where metals might catalyse the development of rancidity in fats. However, in some products, eg some instant desserts, metals such as calcium are required in the gelling process when the product is made, and therefore the use of citric acid will interfere in this process.

Malic and tartaric acids are also used in foods and occur naturally in grapes.

Figure 1.64 Malic acid

CH(OH)COOH
|
CH₂COOH

Figure 1.65 Tartaric acid

CH(OH)COOH
|
CH(OH)COOH

(For cream of tartar, replace one H or a carboxyl group –COOH with potassium K.)

CH(OH)COOH
|
CH(OH)COOK

In unripe grapes malic acid predominates and in ripe grapes there is more tartaric acid. Wines from Germany, where grapes do not always fully ripen, have a higher proportion of malic acid than tartaric. Wines from Spain and warmer regions always have a predominance of tartaric acid. This difference helps to give individual characteristics to wines.

Lactic acid is being used increasingly as a food additive and unlike the previous three acids is in liquid form.

Figure 1.66 Lactic acid

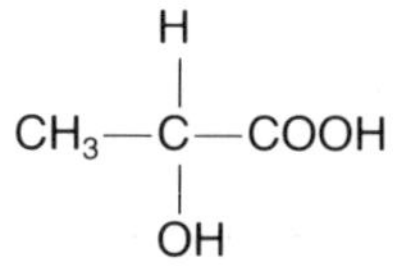

$$CH_3-\underset{\underset{OH}{|}}{\overset{\overset{H}{|}}{C}}-COOH$$

Anticaking agents

These substances have the property of picking up moisture from dried foods without themselves becoming obviously wet. Usually they are anhydrous salts and there are over 20 examples used in food. Common examples are calcium phosphates, magnesium oxide, silicates and salts of some long chain fatty acids such as stearic, palmitic and myristic. Addition of the agents to powders ensure that they retain the free-flowing characteristics, for example, salt.

Buffers

Buffers control and stabilize the pH of a food. However, many food components are natural buffers, eg amino acids and proteins, and any addition of acid or alkaline will have little or no effect. Weak acids and salts of weak acids are used as buffers, such as lactic, fumaric, citric, malic and tartaric acids. Some phosphoric acids also are used.

Firming and crisping agents

These substances are used to prevent loss of moisture and, therefore, texture of vegetables. Salt can act in this role but the best agents are calcium compounds.

Humectants

Humectants act in a manner which is opposite to that of anticaking agents in that they retain moisture and keep a product, such as cake, moist and reduce dehydration. Humectants absorb water and then transfer it to the product as it loses its own moisture. Glycerol has been used widely for this purpose and similarly sodium lactate has found increasing use.

Nutritive additives

A balanced diet should provide all the body's requirements; however, in times of shortage, famine, war and to ensure an adequate intake of nutrients, many foods are fortified with additional nutrients. A number of foods are fortified with vitamins, for example, by law margarine must have vitamins A and D added. Similarly white flour is fortified with calcium and iron (the use of thiamin and nicotinic acid in flour may be abandoned). Breakfast cereals and baby foods are often fortified with vitamins and some minerals, eg iron.

Preservatives

Chemical substances can be added to foods to prevent microbial spoilage, but their use, like all additives, is carefully controlled by food regulations. Antibiotics have been used for preservation of foods but their use is now prohibited. Micro-organisms can become resistant to antibiotics and can pass on their immunity to other, possibly harmful, bacteria. The main preservatives are listed in Table 1.16.

Table 1.16 Presevatives

Benzoic acid	E.210	
	Benzoates:	Ethyl 4-hydroxybenzoate E.214 Methyl 4-hydroxybenzoate E.218 Propyl 4-hydroxybenzoate E.216
Propanoic acid	E.280	
Sodium nitrate Sodium nitrite	E.251 E.250	
Sorbic acid	E.200	
Sulphur dioxide (sulphites)	E.220	(sulphites E221/227)

However there are 35 permitted preservatives in total.

Bacteria are killed by sulphur dioxide, benzoic acid and the benzoates. The use of these compounds is widespread and started with the burning of sulphur to sterilize wine barrels. Sorbic and propionic acids prevent the growth of moulds in cheese, flour, confectionery and bread.
(*NB* The current regulations should be checked for permitted usage.)

Nitrate is converted to nitrite in the curing of meat, so both are permitted preservatives, and nitrite is responsible for the pink colour of cured meats. Nitrite prevents the growth of a very dangerous organism *Clostridium botulinum* which can produce a deadly toxin in many foods under the right conditions. Nitrite can combine with amino compounds to produce nitrosoamines which are carcinogenic. Although nitrite is still permitted because of its usefulness, it should be used in minimal quantities.

Sequestrants

We have already seen that some acids such as citric acid can combine with metal ions and make them unavailable. Similarly, naturally occurring phytic acid combines with calcium. A number of acids will act as sequestrants, such as phosphoric acid and tartaric acid.

Figure 1.67 Citric acid acting as a sequestrant

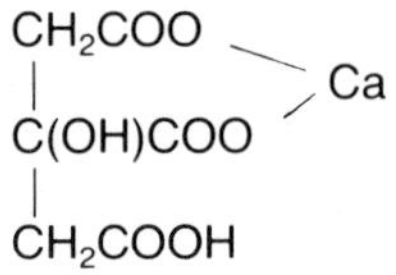

$$
\begin{array}{l}
CH_2COO \\
| \\
C(OH)COO \\
| \\
CH_2COOH
\end{array} \Big\rangle Ca
$$

The most effective sequestrant used to remove metal ions is a synthetic compound EDTA (Ethylene diamine tetraacetic acid) and it is used in food as a calcium salt, particularly in canned fish. Small, sharp crystals of 'struvite' can be found in canned fish and crustaceans. The crystals are ammonium magnesium phosphate and EDTA will prevent their formation by combining with the magnesium.

Review

1. Additives
  – naturally occurring or synthetic substances added to foods
  – used to the advantage of the consumer

2. Acids
  – commonly used in crystal form
  – citric
    malic
    tartaric
  – in liquid form – lactic acid

3. Anticaking agents
  – prevent moisture pick-up by food
  – keep powders free-flowing

4. Buffers – control pH – weak acids and their salts

5. Firming and crisping agents
– control texture in vegetables
– best agents are calcium compounds

6. Humectants
– opposite of anticaking agents
– take up moisture and pass it on to food to prevent drying out

7. Nutritive additives
– to improve nutrients in a food
– some added by law, eg vitamins A and D to margarine

8. Preservatives
– prevent microbial spoilage
– sulphur dioxide and benzoic kill bacteria
– sorbic and propionic acids prevent fungal growth
– nitrite very effective against *Clostridium botulinum*

9. Sequestrants
– bind up metals
– citric, phosphoric, tartaric acids
– EDTA

10. Other additives (covered elsewhere):
– Antioxidants – prevent rancidity in fats
– Colours – attract consumer
– Emulsifiers – emulsification of oil in water
– Enzymes – many uses to change foods
– Flavours – to improve a product or flavour a manufactured product

Practical exercises: *Additives*

1. Salt determination in food

Liquidize the food in a blender and add 10 g to a standard 100 cm^3 flask. Make up to the mark with *deionized water*. Shake well and then filter out solid material. Pipette 10 cm^3 of the filtrate into a flask and titrate with 0·1 M silver nitrate solution. Use fluorescein indicator which gives a pink colour at the end-point. Calculate the salt content of the food as follows:

$$1 \text{ litre of } 0·1 \text{ M nitrate reacts with } (58·5/10) \text{ g of salt}$$

$$1 \text{ cm}^3 \text{ of } 0·1 \text{ M nitrate reacts with } (58·5/10 \times 1000) \text{ g of salt}$$

The volume of silver nitrate used (V) reacts with:

$$V \times (58·5/10 \times 1000) \text{ g of salt}$$

This amount of salt was contained in 10 cm^3 of filtrate, ie 1 g of food.

2. Test for sulphur dioxide (and sulphites)

Soak a filter paper in lead acetate solution (**Poison**).

Place a sample of food in a test-tube, add dilute hydrochloric acid and a small amount of zinc powder. Place the filter paper over the mouth of the tube. The paper becomes blackened when sulphites are present due to the production of hydrogen sulphide gas.

3. Labelling of prepared foods

Examine the labels of a number of prepared dried, canned and frozen foods. Draw up a table of the additives used. If a 'permitted additive' is used try to establish which actual substance is likely to be present. Against each additive, note its particular function in the food, and why it was used. What would be the effect on the food if these additives were not used in each case?

Section 2

Commodities and Raw Materials

In this section of the book individual food commodities will be discussed and knowledge of individual components from Section 1 will be applied to each food, particularly when discussing chemical composition. Principal methods of processing, handling and technological problems will be reviewed. However, each of the commodities covered is in itself a very large subject and therefore only the main areas of interest will be discussed.

2.1 Dairy products

2.1.1 Milk

A perfectly balanced food, hygienically served, at the right temperature, in the right quantities, would be an ideal food for a developing offspring. Milk is the ideal food but with some limitations. All mammals produce milk from specialized glands to feed their young after birth. Each animal produces its milk of a certain composition to meet the needs of its young and not necessarily those of other animals. Cows' milk contains more protein than human milk but less carbohydrate. In simple terms, to convert cows' milk for human consumption it should be diluted with some sugar added.

Milk is also a vehicle by which antibodies can be passed from the mother to the offspring. *The colostrum* is the first secretion of the mammary glands after birth, containing more vitamins, particularly retinol, riboflavin, thiamin and biotin. Colostrum also contains more globular proteins, to which group of proteins antibodies belong. The mother is able to pass on a certain amount of the immunity she has developed to diseases to her offspring.

Milk is unfortunately also a vehicle by which disease can be passed from an animal to its young or in the case of cows' milk to a large number of people. Before pasteurisation was adopted milk from infected cattle

Table 2.1   Composition of milk

Animal	% Total solids	% Fat	% Protein	% Casein	% Lactose	% Ash
Human	—	2·0–6·0	0·7–2·0	—	6·0–7·5	—
Cow	12·6	3·8	3·3	2·8	4·7	0·7
Goat	13·2	4·2	3·7	2·8	4·5	0·8
Sheep	17·0	5·3	6·3	4·6	4·6	0·8

could cause tuberculosis, staphylococcal and salmonella infections, typhoid, paratyphoid, diphtheria and many other diseases.

2.1.1.1 Composition of milk

There are over 100 different compounds in milk, which can be considered as a mixture of carbohydrates, proteins, lipids and many inorganic and organic salts dissolved in water. In Table 2.1 a comparison is made of a number of different milks; figures given may vary somewhat for the reasons which will be discussed later.

Cows' milk is the most widespread milk used throughout the world, but it is legally considered to be genuine only if it contains at least 3% fat and 8·5% other solids.

The composition of cows' milk varies according to breed, age, stage of lactation, season, feed, time and period between milkings. The most important single factor governing milk composition is the breed of cow and the principal breeds in Europe are Friesian shorthorn, Ayrshire, Brown Swiss, Jersey and Guernsey. Some of the differences in milk due to breed are given in Table 2.2. In addition to the factors listed above, two other factors will affect milk composition in many cases and these are disease and temperature. In warmer climates, around 28–32°C, milk yield falls enormously and its composition may alter. Daily injections of the growth hormone, bovine somatotropin, will greatly increase a cow's milk yield. However, this is still not officially licensed by the Government.

Table 2.2   Variation in milk composition due to breed

Breed	% Fat	% Protein	% Lactose
Ayrshire	4·0	3·6	4·7
Friesian	3·9	3·3	5·0
Holstein	3·4	3·3	4·9
Jersey	5·4	3·9	4·9

The *lipid content* of milk is mainly fat (butterfat) with small amounts of phospholipids, sterols, carotenoids and vitamins A and D. The fat occurs in globules which are stabilized by the milk fat globule membrane, a complex structure of proteins, phospholipids and enzymes. Any heat processing will damage this stabilization of the milk fat by denaturing the proteins. Fat globules will then coalesce and rise to the surface to produce the 'cream layer'. Fatty acids in the fat are unusual in being mainly saturated with some very short chain acids such as butyric (butanoic), caproic, caprillic and capric. Polyunsaturated acids, particularly linoleic and linolenic acids are very much in the minority.

Milk proteins are numerous, but are generally classed into two main groups: casein (which is a curd precipitated by acid or the enzyme rennin); and whey proteins.

Casein is a mixture of phosphoproteins (ie proteins containing phosphate groups) which comprise about 80% of all the protein in milk and so total some 2.5–3.2% of whole milk. The different caseins comprising whole casein have different properties and these include α_s which is coagulated by calcium ions and κ-casein which is insensitive to calcium. In milk the casein exists as colloidally dispersed bodies known as *micelles*. The structure of these micelles is still uncertain but probably the calcium sensitive casein ie α_s is on the inside and is protected by calcium insensitive casein ie κ- on the outside of the micelle. This protection of the micelle by κ-casein is destroyed by the action of rennin, which then allows calcium to react with the α_s casein, thus causing precipitation and formation of a curd, an essential preliminary in cheese making. The proteins remaining after casein has been precipitated from milk are collectively known as the *whey proteins* and include albumins, globulins, enzymes, and protein-breakdown products. The most important whey protein is β-lactoglobulin which can be denatured by heat over a period of time.

The protein chains can uncoil during heating, and associate with κ-casein, but in so doing expose certain side groups on the chains, particularly those which contain sulphur. This leads to the typical cooked milk flavours of sterilized and evaporated milk. The other main whey protein, α-lactalbumin, is associated with enzymes responsible for lactose synthesis.

2.1.1.2 Milk processing

As we have seen milk is an excellent food, but it is also excellent growth medium for a vast range of micro-organisms, including a number of pathogenic bacteria. Tuberculosis was widespread until only a few decades ago and milk was the main carrier of the disease. Until a few years ago cattle could be infected with *Brucella*, which, although causing contagious abortion in cattle, causes an undulant or repetitive, 'flu'-like disease in humans. Milk can carry this disease and many others have been reported, including typhoid. Bacteria can enter the milk from the udder, operatives during milking, utensils and equipment. Heat treatment of milk is therefore essential to remove harmful bacteria and those which will cause rapid spoilage of the milk.

Pasteurization, developed by Louis Pasteur whilst working on spoiled wine, consists of heating milk below the boiling point but at a temperature high enough to kill harmful organisms and to reduce spoilage organisms (about 99% are killed). The time and temperature of pasteurization is fixed to kill the organism *Mycobacterium tuberculosis,* the causative organism of TB. Milk used to be pasteurized in bulk in the 'holder process' when about 300 gallons would be heated and stored for 30 minutes at 62·8–65·6°C (145–150°F). All batch processes in food production are less efficient, usually demanding more manpower, and less cost-effective than continuous processes. The high-temperature short-time (HTST) process is a continuous process which was developed to offset the problems of the holder process. If a higher temperature is used for pasteurization the kill of bacteria can be achieved in a much shorter time and damage to the flavour and nutritive value of the milk will be less.

A temperature of 71·7°C (161°F) is used for 15 seconds to achieve HTST

Figure 2.1 HTST pasteurization of milk

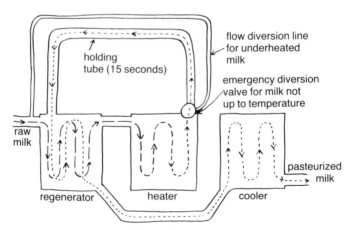

On entering the unit, raw milk is first warmed by already pasteurized milk in the regenerator. It is heated to pasteurization temperature in the heater and then passes to the holding tube. Insufficiently heated milk is diverted by thermostatically controlled valves and is reprocessed. After holding for 15 seconds the milk passes to the regenerator to warm incoming milk and then to the cooler. The unit is built up of plate heat exchangers and milk does not come into contact with steam or cooling water.

pasteurization. Special plate heat-exchangers, with milk on one side and steam on the other, are used with a special holding tube through which the milk passes to the cooling section and hence to bottling.

Pasteurization of milk has very little damaging effect with the exception of the loss of about half the vitamin C, and about 10% of thiamin, and cyanocobalamin (B_{12}) which reacts with breakdown products of vitamin C. Protein of the membrane surrounding the milk fat globules is slightly denatured and this allows fat globules to agglomerate and rise to the surface to form the characteristic cream layer. Leaving bottles of milk out in sunlight does far more damage in causing the destruction of all the remaining vitamin C and the production of off-flavours.

The process of *homogenisation* is sometimes carried out to prevent the separation of the cream layer and is essential when milk is subjected to higher temperatures in sterilization processes. To homogenise milk it must firstly be warmed to about 60°C and then forced through a small gap in a pressure homogeniser which reduces the fat droplets to about 1–2 mμ (microns ie 0·001–0·002 mm). The small fat droplets are very stable and do not separate out into a cream layer.

Sterilization is a much more severe heat process destroying all micro-organisms and most of their spores. Sterilized milk, in its typical crown-capped bottle, was developed in the Midlands so it could be left on the doorstep all day while the family was at work. The strong cooked milk flavour is still preferred by many people. The traditional method of producing sterilized milk is by in-bottle sterilization in which milk is filled and sealed into the bottles, then heated to 120°C (249°F) for 15 minutes. The bottles are cooled by water spray to about 75°C, then allowed to cool slowly to room temperature.

The flavour of sterilized milk takes some time to develop, and therefore it is possible to heat milk to a high temperature to sterilize it but in a shorter space of time to prevent flavour development. In this way the milk is similar to pasteurized milk. The *UHT (ultra-high temperature) process* was developed for this purpose and involves heating milk in a plate heat-exchange unit at 132°C (270°F) for 1 second. 'UHT' must be homogenised and it is often packed in cardboard cartons and sold as 'long-life milk'.

A more recent method of sterilization, used extensively in Europe, is called *Uperization* and involves injection of steam under pressure into the milk to obtain rapid sterilization. UHT sterilization of milk has little effect on nutritive value, in a similar manner to pasteurization. However, dissolved oxygen in the milk may cause severe losses in vitamins C, B_{12} and folic acid over a period of time. The injection or uperization method causes removal of dissolved oxygen. In contrast to the UHT method the in-bottle sterilization process causes an overall reduction in the nutritive value of milk. The vitamins are badly affected in that vitamin C content is halved, B_{12} is destroyed and one third of thiamin is lost. The biological value of the proteins is reduced.

2.1.2 Cream

Cream is milk in which the fat content has been greatly increased by separating out some water. A special cream separator, which is like a centrifuge built up of a number of plates or cones, is used to remove water by centrifugal force as water is heavier than butter fat. The separator operates at about 6500 rpm.

Figure 2.2 A cream separator

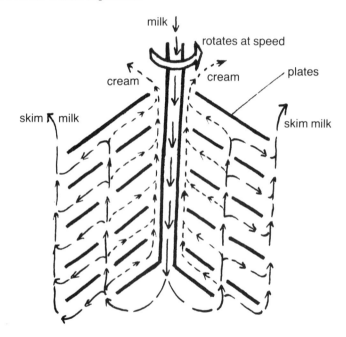

Table 2.3 Types of cream

Type	Fat	Protein	Lactose	Water
Single	18%	3%	4%	74%
Double	48%	1·5%	2·5%	48%
Whipping	35%	2·2%	3%	59%
Clotted	55%	3%	2·5%	40%

By this method three types of cream are produced, and these are given in Table 2.3, together with clotted cream.

To whip cream into a stable form the fat content must be between about 30 and 38%. A fat content of 36% produces a good foam in which air bubbles are trapped and the cream increases in volume up to three times. This increase in volume is called 'over-run'. *Clotted cream* has the highest fat content and is made by a slow traditional method. Milk is firstly allowed to stand in shallow containers for almost 12 hours at room temperature and then it is heated to about 80°C and cooled again slowly for about a day. The cream separates out into a thick layer and is scraped off the top of the milk and packed.

2.1.3 Butter

It is appropriate to consider butter after discussing cream, as it is made by churning cream. The churning process causes what is known as an 'inversion of the colloid'. Cream is an emulsion of fat in water, whereas butter is an emulsion (or colloid) of water in fat, which is obviously an inversion of the original colloidal state and involves the expulsion of much of the water from the original cream. Thus cream of 35–38% fat and 60–65% water is converted to butter of 80% fat and 16% water (this being the legal maximum for water content).

Fresh cream, separated prior to churning, is used extensively for butter making, but it is also possible to ripen or sour cream to give added flavour. Special cultures of organisms are added to the cream and are allowed to grow to produce about 0·25% lactic acid. Lactic acid producing bacteria are used, such as *Streptococcus lactis* or *cremoris*. Butter can be salted with up to 2% salt added, but unsalted butter is now more popular.

Before churning the cream is pasteurized and cooled to about 10°C (50°F) at which temperature churning is most efficient. Churning is effected by violent agitation of the cream, which takes about 30 minutes. Small pieces of butter at first appear and then the 'butter-milk' is removed. The butter is washed, salted and 'worked' to obtain the desired consistency and a final water content of up to 16%.

The nutritive value of butter is somewhat variable, as it depends on its fat content and fat-soluble vitamins, mainly retinol (A) and calciferol (D). Carotenoids are the reason for the yellow colour of butter and in summer, because the cow has plenty of green grass, the butterfat is richer in carotenoids and a brighter yellow.

2.1.4 Margarine*

A substitute for butter which was cheap and readily available was sought in the last century and Napoleon III awarded a prize to Mèges Mouriès in 1869 for inventing the first margarine. Margarine, like butter, is an emulsion of an aqueous phase of milk and water dispersed in a fatty phase, usually made of vegetable oils. The original margarine contained beef tallow and now any blend of fat or oil can be used to give the same margarine after processing. Hard margarines use hydrogenated fats but soft-spread margarines use oils which are polyunsaturated and usually low in cholesterol. Examples of three different fat blends used for margarine are given in Table 2.4; however there are numerous other combinations possible.

Table 2.4   Fat blends for margarine

Fat	% of fat mixture
1. Coconut oil	55
Hydrogenated vegetable oil	20
Groundnut oil	25
2. Hydrogenated palm kernel oil	70
Coconut oil	20
Palm kernel oil	10
3. Coconut oil ⎫ Palm kernel oil ⎭	40*
Palm oil	7
Hydrogenated groundnut oil	13
Hydrogenated whale oil	20
Groundnut oil	20

*(1939–45)

The carefully blended oils and fats must be odour-free; have a bland taste; and above all have a wide melting range to simulate the sensation of butter melting in the mouth. Skimmed milk, to which has been added a starter culture of bacteria, is allowed to sour to produce a butter-type flavour (diacetyl).

Salt, vitamins A and D, colour and sometimes flavour are added to the skimmed milk. The oil blend and the skimmed milk preparation are blended together in the right proportions in a large vessel which has rotating paddles. An emulsion is gradually formed and is aided by the addition of emulsifying agents such as glyceryl monostearate and lecithin.

* Not a true dairy product.

The emulsion formed so far in the process is not margarine, as it still does not possess the right texture. The emulsion is passed to a machine called a 'votator' which is in fact a closed cylinder within another cylinder. The inner cylinder is cooled and in coming in contact with this the margarine solidifies and is worked into a semi-solid condition which can then be packed.

The greatest challenge facing margarine manufacturers has been to simulate the flavour and mouthfeel of butter. In the past margarine was always inferior to butter. However, the controversy of hard animal fats has resulted in rapid growth of margarine sales, particularly of soft, low in cholesterol, margarines. The vitamin content is controlled by law at 900 μg of vitamin A and 8 μg of vitamin D per 100 g of margarine, thus making it better than most butters as a source of these vitamins. Fortunately for margarine manufacturers, the days of the product being the poor man's butter have passed and we are now in a time of over-production of butter and the accumulation of butter 'mountains'.

2.1.5 Skim milk

Milk from which almost all the fat has been removed is known as skim milk. However, a small fraction of the fat, perhaps about 0·1%, is almost impossible to remove by the standard method using a centrifugal separator (Figure 2.2). Removing the fat from milk obviously removes the fat-soluble vitamins, and so skim milk must not be used to feed babies. Because of the controversy surrounding butterfat and its connection with the blocking of arteries, skim milk is becoming more popular and more readily available. Most skim milk is sold in powder form.

2.1.6 Dried milks

As milk is about 87% water, in order to transport a large amount of milk great distances it would be advantageous to reduce the bulk of this water or completely remove it. Drying removes most of the water from milk and also is an excellent means of preservation. The process can produce an excellent product which when reconstituted differs little from fresh milk.

There is little loss of nutritive value of milk during drying if it is carried out correctly. Overheating during drying will cause significant vitamin losses, protein damage and will encourage browning by non-enzymic means, such as the Maillard reaction. The lactose, a reducing sugar, and the amino acid lysine, an essential amino acid, have been implicated in this browning. When dry, milk powders must be kept from contact with moisture and ideally from air. If the moisture level of a powder is allowed to rise, to about 5%, the Maillard reaction will slowly take place turning the powder from white, through cream to light brown. This can be observed sometimes in a tin of milk powder which is used infrequently over a long period. The fat in *whole milk powder* is liable to undergo oxidative rancidity fairly rapidly and therefore must be packed in the absence of oxygen. Its shelf-life under ideal conditions is significantly less than skim milk powder. A compromise product, *filled milk,* has been produced for a number of years, particularly in the war and is still popular in the catering trade. Filled milk is skim milk to which is added vegetable oil, it is then homogenised and dried. Various levels of fat in the product are possible and different fats can be used according to specific uses. Filled milk does not offer the possible health risks associated with butterfat and keeps much better than whole milk powders.

Most milk powders are very fine powders which, when added to water, tend to float on the surface, have poor *dispersability* and poor *wettability*. A process of *instantization* can be used in which the powder is slightly rewetted so it clumps together. These clumps of powder act like sponges and absorb the water and disperse in it rapidly.

Traditionally, but rare nowadays, milk powder was made by *roller drying*. The roller drier (see Section 4.3.1) consists of a hollow drum, internally heated by steam, to which a film of milk adheres, and, as the drum rotates, the milk dries to give a flaky powder. The method produces an almost sterile product as the heat treatment during drying is severe, but also the protein may be damaged and will not reconstitute in water easily. Baby milk products were made by this process.

Spray drying (see Section 4.3.1) has replaced roller drying as it produces a product which is more soluble, of better flavour and colour. Milk is concentrated in an evaporator and then sprayed whilst still hot (80°C/176°F) into a chamber where the spray meets a blast of hot air (180°C/388°F) and dries instantly.

2.1.7 Concentrated milks

Water can readily be removed from milk by evaporation. Products were made for many years by this method, but all had a marked 'cooked milk flavour'. The advent of the evaporator working under a strong vacuum enabled the concentration to be undertaken at a lower temperature, thus reducing the heat damage, for example at 50°C/122°F. *Evaporated milk* is made by evaporating water from milk to reduce the water content to about 70%. The product sometimes has a poor granular texture which can be improved by adding either sodium citrate, disodium phosphate or calcium chloride. The product is still highly perishable and has to be homogenised, then sealed into cans, and heat processed in retorts at 121°C/250°F for about 15 minutes, depending on the can size. For many years evaporated milk was popular as a 'cream-substitute' to add to fruit dishes. *Condensed milk* is a similar product to evaporated milk, but is sweetened, and in fact depends on its sugar content for preservation. Milk must firstly be pasteurized, as the product is not heat processed, but is not cooled as it passes directly to the evaporator. A sugar syrup (60–65% sucrose) is added to the evaporator and the whole milk is concentrated under vacuum at 50–55°C (121–127°F). The product is then cooled and agitated at the same time. Often very small crystals of lactose are added to ensure rapid crystallization of the lactose. The addition of the large amount of sucrose forces lactose, less soluble than sucrose, out of solution. If the lactose is allowed to crystallize slowly it produces large crystals, sometimes as big as marbles. Agitation and seeding with small crystals ensures small lactose crystals which are unnoticed in the product. The product is poured into sterilized cans and sealed without further heat treatment. Condensed milk depends on its high sugar content for its preservation and therefore careful hygiene is needed during this process. Again, this is a product of yesteryear.

2.1.8 Ice cream

A type of ice cream originated in Paris in 1774, but frozen desserts of various types have been known since Roman times. Ice cream is an emulsion of fat in a complicated solution which is made up of both colloidal and true solutions. Ice cream contains very small crystals of ice, air sacs, fat globules, colloidal suspensions of casein, stabilizing agents, flavours, colour and sugar solution.

Commercially there are two main types: soft ice cream and hard ice cream.

Soft ice cream

This type of ice cream has grown in popularity over the last few years and tends to be made locally in small batches. The secret in developing an ice cream mix is to balance the ingredients correctly; too much milk powder, for example, leads to sandiness as lactose may crystallize from the powder. Too little milk powder, which tends to absorb about seven times its own weight of water, will mean too much water, which will form large coarse ice crystals. Stabilizers and emulsifiers are used extensively in ice cream and these include gelatin, alginates, modified celluloses, carageenans, pectins and various gums such as gum tragacanth.

These substances prevent the formation of large ice crystals during the freezing operation and allow only small crystals to form. Stabilizers give body to the product and improve the melting resistance of the ice cream. The basic principle of operation of a stabilizer is to absorb a large quantity of water and without them ice cream has poor texture and rapidly melts. A simple formulation for a soft ice cream is given in Table 2.5.

The ingredients are mixed together and are usually homogenised then pasteurized. Air is whipped into the ice cream as it is frozen, usually at about −5°C/22°F. Compared with hard ice cream there is less air mixed into the product and so it only increases in volume by about 50%, ie an overrun of 50%. The product is usually served directly from the freezer.

Hard ice cream

For a hard ice cream at least 8% fat should be used, but in some processes if the figure goes much above 9% the ice cream may be very heavy, as it will be impossible to mix in enough air to give sufficient overrun. A simplified formulation is given in Table 2.6. In both types of ice cream the sugar content should not exceed 15%, as it may crystallize out, but small amounts tend to encourage the growth of large ice crystals.

Table 2.5   A soft ice cream formulation

	%
Skim milk powder	11
Fat (vegetable)	6
Sugar	14
Stabilizer	1
Water	68
(Flavouring and colour as required)	

Table 2.6   A hard ice cream formulation

	%
Skim milk powder	10
Fat (vegetable)	
(dairy ice cream − butterfat)	12
Sugar	14
Stabilizer	1
Water	63
(Flavouring and colour as required)	

NB Legally ice cream must contain at least 7·5% solid-not-fat and at least 5% fat.

The water used must be warm to aid dissolution and dispersion of the ingredients. The mix is homogenised and pasteurized, then frozen. During the freezing operation the mix is stirred vigorously to incorporate more air than in soft ice cream, otherwise the mix would freeze to a hard mass. An overrun of about 100% is essential in hard ice cream. Once frozen the ice cream may be cut into blocks and wrapped, then hardened at $-40°C/-40°F$.

Ice cream has an excellent 'health record', having been involved in few food poisoning outbreaks. It is an energy-supplying food because of its sugar content, but only contains about 3·5% protein.

2.1.9 Cheese

Cheese has been a food from the earliest times and has been made from most milks, particularly goats', sheeps', buffalos' and of course cows'. There are several hundred varieties of cheese and it is surprising that a product, depending on complex biological reactions, is extremely consistent in appearance, flavour and nutritional qualities.

Cheese production depends on a number of biological reactions which must occur in the right sequence and which can easily be affected by a number of interfering factors.

In 1991 research work has indicated that a small amount of cheese in a meal helps repair teeth.

Cheddar cheese manufacture

Pasteurized milk is placed in a large stainless-steel vat equipped with an outer water-jacket. A special culture of lactic acid forming bacteria is added, together with colouring. Usually a culture of *Streptococcus cremoris* and *Streptococcus lactis* is used. Unfortunately, starter cultures can be attacked by viruses known as *bacteriophages* which kill the lactic acid producing bacteria. Special care is always taken to prevent 'phage attack as the starter will be inactivated and a cheese of the right acidity and texture will be impossible to produce. The milk is held at slightly above room temperature (25°C/86°F) and lactic acid is produced to about 0·17–0·2%.

At this stage of acidity the *rennet*, a preparation of the enzyme rennin (or chymosin), is added to coagulate the protein in about 20 minutes. Let us look again at the casein micelle, Section 2.1.1.1, in which the calcium sensitive $α_s$-casein is protected by the κ-casein which is insensitive to calcium. The rennet attacks the κ-casein and makes it insoluble.

Thus the $α_s$-casein is no longer protected from calcium by the κ-casein and so the $α_s$-fraction combines with calcium and gels to form the *curd*. The curd coagulates into a solid mass which is cut into cubes, either by

115

Figure 2.3 Traditional cheese cutters

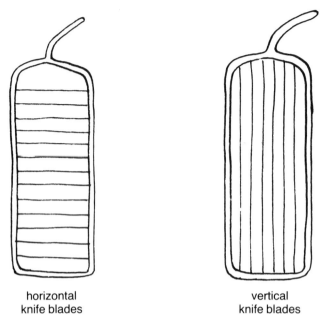

horizontal	vertical
knife blades	knife blades

traditional cutters as shown in Figure 2.3, or by more modern automatic systems.

The temperature of the vat is now increased to about 40°C/104°F to make the curd contract and expel the liquid whey, which is now drained away.

The curd cubes coalesce and settle into a firm mass which is cut into slabs about 30 cm by 60 cm. The actual process called 'cheddaring' now begins and involves piling up the slabs of curd and turning them every 15 minutes for about 2 hours. During this process subtle changes occur in the milk proteins so that the desired firmness and texture of cheese begins to develop.

The slabs are milled into small pieces and the last drops of whey are allowed to escape. Salt is added at this stage at about 0·1 to 0·2% and mixed into the mass of curd-cum-cheese. These pieces are filled into moulds and pressed for about 24–48 hours. The moulded cheeses are covered in cheese-cloth or special plastic coverings. At this stage the cheese is termed a 'green cheese'. The cheese is stored at 10°C (49°F) for about 3 months to ripen or cure.

Ripening is the period during which the cheese develops its own special character. Enzymes, bacteria and moulds all play a part in ripening the cheese, and the changes which occur are numerous. There are perhaps three general reactions which occur: protein is hydrolysed by proteolytic enzymes; lipids are attacked by lipases; and amino acids and fatty acids form flavour compounds. If most of the proteins are hydrolysed the cheese will become very soft and creamy. Amino acids will be liberated from the protein chains and will contribute to the flavour of the cheese. The breakdown of the lipids, which contain short chain fatty acids, contributes significantly to flavour of cheese. Effectively the process of hydrolytic rancidity is followed in that free fatty acids, such butyric (butanoic), are produced and have strong flavours. Amino and fatty acids may also be attacked by enzymes to release flavour compounds such as ammonia.

Other cheeses

There are numerous semi-hard and hard cheeses made by modified methods of the Cheddar process, and many are given place names, eg Derby, Cheshire and Gloucester. Each variety is produced to give a characteristic level of acidity and moisture which allows the development of the desired flavour and consistency.

Harder cheeses are produced by: heating the curd to a higher temperature; cutting the curd finer during milling; and by applying a higher pressure during the pressing into moulds.

Blue cheeses are not pressed but gently packed into moulds so that *Penicillium* blue mould can grow through the cheese lumps. The most common mould is *P. roquefortii*. The mould produces proteolytic enzymes and often ammonia is released which is a characteristic of the flavour of some of these cheeses.

The *soft cheeses*, such as Camembert and Brie, depend on a different mould *Penicillium camembertii* which produces a white growth. The curd is put into shallow moulds and rubbed with salt which allows the mould to grow. Enzymes from the mould act on the curd to produce the soft and creamy texture of the ripened cheese. Once ripe the shelf-life of these cheeses is short.

Processed cheese is a manufactured product from a number of cheeses. It is made by emulsifying the cheese with green cheese in the presence of emulsifier and water. The emulsifying agents used are blends of salts: sodium and potassium phosphates and calcium, potassium or sodium citrates.

The cheese is chopped into small pieces, heated and mixed thoroughly with green cheese, water and emulsifier. It is often wrapped in metal foil, and as the heating process kills many organisms it remains moist and keeps for some time.

Cheese *spreads* contain more moisture and often have gums or gelatin to help form a smooth paste.

2.1.10 Fermented milks

Yoghurt is the best known fermented milk in most Western countries, although it originated from the Balkans and Middle East; but there are a large number of other products such as: cultured buttermilk, filmjolk, kefir, kumiss and laben. Although yoghurt has been made in the U.K. since the First World War it is only during the last 15 years that sales have increased rapidly, particularly after the advent of fruit yoghurts.

Yoghurt is milk, often concentrated or with added milk powder, which has developed a characteristic acidity and flavour due to the growth of two micro-organisms, *Lactobacillus bulgaricus* and *Streptococcus thermophilus*. The two organisms must be in equal amounts and one must not outgrow the other or a bitter or too acid a product will result. Yoghurt has long been thought to have therapeutic properties, particularly for convalescents. Most yoghurts do not have such properties, but there are obviously some good nutritional aspects to be considered as yoghurts contain more protein, thiamin, and riboflavin than milk, but low fat yoghurts contain less fat-soluble vitamins. However, if the bacterium *Lactobacillus acidophilus* is included in the starter culture then the yoghurt may help patients to recover from gastroenteritis and other gastric disorders. Yoghurt falls into main types according to its consistency, set and stirred yoghurt. To make *set yoghurt* the fermentation is allowed to take place in the container in which the yoghurt is sold, whereas *stirred yoghurt* is fermented in bulk then packed later.

The ingredients for yoghurts include whole milk, skim milk, evaporated milk, dried milk, stabilizers and thickeners, fruit flavours, colours and sugar. Generally, low fat yoghurts are produced so the fat is separated from the milk using a centrifugal separator. The solids content of the milk is increased by: evaporation of some water, or addition of evaporated milk, or addition of skim milk powder. Ingredients such as sugar, stabilizers, colours and flavours are blended into the milk base. The mix is homogenised then pasteurized at 90°C/194°F for 30 minutes, or by a HTST method, to kill all micro-organisms. The mix is cooled to 44°C/110°F and inoculated with the starter culture.

To make a *set yoghurt* the mix is incubated at 44°C/110°F for about 1½ hours, then poured into containers which are kept warm until the yoghurt has fully coagulated. The yoghurt is then cooled to 5–8°C/41–46°F and held at this temperature until consumed, ideally within 14 days. Natural yoghurts are usually of the set type.

To make *stirred yoghurt,* which offers advantages in continuous manufacture, incubation is at a slightly lower temperature so that the yoghurt becomes thicker but not coagulated, and continuous stirring ensures that no curd is formed. Fruit and syrup are metered into the containers, followed by the yoghurt mix, which is then cooled and stored as before.

New developments in yoghurts include frozen yoghurt and long-life yoghurt where aseptic packaging is used following heat treatment of the yoghurt. Paradoxically, during 1987 there was a rapid growth in sales of Greek-style yoghurts containing up to 10% fat.

Review

1. Milk
  – produced by mammals after birth of young
  – ideally balanced food for that particular animal's young
  – human milk more lactose but less protein than cows' milk
  – colostrum first secretion after birth, rich in antibodies and vitamins
  – milk vehicle by which diseases may be passed from cow; eg TB, typhoid, diphtheria and salmonellosis

2. Cows' milk composition
  – milk legally genuine if at least 3% fat and 8·5% solids-not-fat
  – varies according to breed, age, stage of lactation, season, feed, time and period between milkings
  – Guernsey and Jersey richer in fat
  – lipid content high in saturated short chain fatty acids, eg butanoic (butyric) acid
  – caseins – phosphoproteins – about 80% of all milk protein
  – α_s-casein sensitive to calcium
  – κ-casein *in*sensitive to calcium
  – whey proteins – β-lactoglobulin when denatured gives 'condensed milk' flavour
  – α-lactalbumin

3. Milk processing
  – pasteurization 71·7°C for 15 seconds
  – to kill pathogens, eg *Mycobacterium tuberculosis* and most spoilage organisms
  – homogenisation breaks down fat globules and prevents cream-layer forming
  – sterilization – 'in-bottle' traditional process, heat to 120°C for 15 minutes
  – 'cooked milk' flavour obvious
  – UHT (Long Life) sterilization for 1 second
  – no flavour change
  – uperization, sterilization by direct steam injection

4. Cream
  – separated by centrifugal separator
  – single cream – 18% fat
  – double cream – 48% fat
  – whipping cream – 35% fat
  – clotted cream – 55% fat

5. Butter

– made by churning cream
– 'inversion of colloid' as cream
    oil in water emulsion changed to water in oil emulsion in butter
– butter 80% fat, maximum 16% water
– starter culture may be used of *Streptococcus lactis* or *cremoris*
– up to 2% salt may be added

6. Margarine

– butter substitute, but now nutritionally superior and less a health-risk
– hard margarines use hydrogenated vegetable fat
– different blends of oils and fats used to give same product
– skim milk preparation to give butter flavour blended and emulsified with fat
– Vitamins A and D added by law

7. Skim milk

– milk after removing fat
– no risk from saturated fatty acids
– more readily available but most dried

8. Dried milks

– mostly skim milk, but some whole milk powders
– fat in whole milk may become rancid – air must be excluded
– moisture uptake of powder may cause Maillard reaction
– filled milk – vegetable fat added to skim milk
– fine particle powders hence poor dispersability, wettability and hence solubility in water
– powders rewetted to form clumps which act as sponges so disperse easily in water
– most powders spray dried

9. Concentrated milks

– evaporated milk – 'cooked milk flavour'
– salts added, eg sodium citrate to improve consistency
– needs to be heat processed in can
– condensed milk (sweetened) – large amount of sugar added, no necessity to heat process in can
– must pasteurize then concentrate by evaporation
– lactose forced out of solution by sucrose

10. Ice cream

– emulsion of fat in complicated solution
– very small ice crystals
– soft ice cream – milk powder, less fat than hard ice cream, stabilizer – freeze at −5°C, serve from freezer
– hard ice cream – at least 8% fat – sugar not more than 15%
    – freeze and whip in air to give 100% overrun
    – harden at −40°C
– legally ice cream must contain at least 5% fat and 7·5% solids-not-fat

11. Cheese

- depends on producing lactic acid in milk and coagulating protein with rennet
- culture of *Streptococcus cremoris* and *lactis*
- culture may be attacked by bacteriophage
- lactic acid produced to 0·17–0·2%
- rennet attacks κ-casein to make it insoluble
  - calcium can then combine with α_s-casein and precipitate it to form curd as basis to cheese
- ripening of cheese involves:
  - (1) hydrolysis of proteins
  - (2) hydrolysis of lipids
  - (3) conversion of amino acids and fatty acids to flavour compounds
- 'Blue cheeses' – have blue moulds eg *Penicillium roquefortii* to produce colour and flavour
- 'Processed cheese' – emulsifying cheese with green cheese, emulsifier and water

12. Fermented milks

- yoghurts, two types: set or stirred
- starter, to produce acid and flavour of *Lactobacillus bulgarius* and *Streptococcus thermophilus*

Practical exercises: *Dairy products*

1. Microscopic examination

Using a small quantity, thinly spread out, make slides of milk, homogenised milk, cream and skimmed milk. Observe the fat droplets and compare their size in each product.

2. To test milk for minerals

Warm about 100 cm^3 of milk to 40°C and add several drops of glacial acetic acid (CARE!). Stir, and continue to add acid until flocculation occurs. Filter, then test the filtrate for chloride, nitrate, sulphate and phosphate (see minerals section for details).

3. Titratable acidity of milk (in terms of lactic acid)

Pipette 25 cm^3 of milk into a flask, add phenolphthalein as an indicator and titrate with 0·1 M sodium hydroxide. The end-point is a pale pink colour.

1 cm^3 of 0·1 M NaOH is neutralised by 0·009 g of lactic acid

∴ g of lactic acid per 100 cm^3 of milk will be 0·009×titre×(100/25)

(*NB*: titre = volume of 1·0 M NaOH used.)

4. Butter manufacture

Allow 3 pints of milk to stand in a refrigerator for two days. Carefully remove the cream layer from each. Pour the cream into a large stoppered container and shake vigorously for some time. The cream will eventually break and butter will form. Carefully remove the butter milk to leave the butter. A small amount of cold water should be added to wash the butter. Remove surplus water and add a small amount of salt. Taste and compare with commercial butters.

5. Clotting of milk with rennet

Special requirement: commercial preparation of rennet. Keep in a refrigerator.

(a) Effect of rennet concentration:
To 10 cm^3 of milk in each of four test-tubes add:
  (i)   0·5 cm^3 of a 5% solution of rennet in water
  (ii) 1·0 cm^3 of a 5% solution of rennet in water
  (iii) 2·0 cm^3 of a 5% solution of rennet in water
Place in a water-bath at 30–35°C and note the time curd starts to appear in the milk.

(b) The effect of temperature:
Repeat (a) at temperatures of 20°C, 40°C and 50°C.

(c) The effect of pH:
To four test-tubes containing 10 cm^3 of milk each add:
  (i)   0·5 cm^3 of lactic acid solution (1%)
  (ii) 1·0 cm^3 of lactic acid solution (1%)
  (iii) 1·0 cm^3 of 0·1 M sodium hydroxide
  (iv) 2·0 cm^3 of 0·1 M sodium hydroxide
Place in a water-bath at 30–35°C and add 0·5 cm^3 of a 5% solution of rennet. Note time of clotting of each sample.

6. Preparation of casein from milk

Use a 25 cm^3 measuring cylinder and 5 cm^3 of milk and 5 cm^3 of water; place in a water-bath at about 35°C. Add 0·5 cm^3 of acetic acid (10%) to precipitate the casein. Then add 0·5 cm^3 of 0·1 M sodium acetate to buffer the pH at 4·6. Cool and allow to stand. Pour off whey proteins and other liquid to leave casein.

7. Yoghurt – to investigate the effect of temperature on yoghurt production.

Heat, but do not boil, about ¾ pint of milk. Cool to 40°C. Mix six teaspoonfuls of plain yoghurt (starter culture) with a little of the milk, then add this preparation to the remaining milk. Divide into three samples and seal into suitable containers. Store the containers in a refrigerator, in a room and near a boiler. Note the temperature and examine the samples for curd formation, noting the time when it firstly appears. (Samples of yoghurt produced should not be eaten as there may be a risk of contamination with harmful organisms.)

2.2 Meat, fish, poultry and eggs

2.2.1 Meat

Meat could simply be defined as the flesh of animals used as food. To be more precise it normally refers to muscle, associated connective tissue and adjoining fat. Offal such as liver, kidneys, heart and brains may also be referred to as meat. However, the flesh of an animal immediately after death is not meat as a considerable number of biochemical changes are necessary to produce meat of the right colour, texture, flavour and cookability.

Meat can supply most of the requirements of man for growth and normal health, as it is a source of all the essential amino acids, many vitamins, essential fatty acids and minerals. Sometimes meat is of poor digestability as it contains too much connective tissue and this fault is made worse by poor cooking. Meat of this type may remain too long in the digestive system and there is some evidence that this may be associated with certain intestinal diseases. There is an obvious parallel here with the symptoms associated with lack of roughage in the diet.

Meat is considered to be a main source of protein, but protein can be produced more quickly and much cheaper by other means, for example, low fat soya flour contains 50% whereas lean beef only 25%. Single cell protein from micro-organisms is a very efficient system, but there still remains the problem of acceptability. The high cost of meat has led to a decline in sales during the last decade, only beef shows a resistance to the trend in being the most popular meat. Beef has declined by about 10%, whereas lamb and ham have declined by 20%. These figures have been matched by an increase in the consumption of chicken (up 25%) and pork (up 5%).

The brain disease Bovine spongiform encephalopathy (BSE) has affected meat sales, but no evidence exists that the disease can be passed to humans.

2.2.1.1 Muscle structure

The muscle of animals, which converts to meat, is *striated* or *voluntary muscle,* and consists of long cylindrical cells, *the muscle fibres,* which are parallel to each other. There are striations going across the muscle cells, hence the name *striated muscle.* The muscle fibres are held in bundles by connective tissue, generally small bundles of cells give more tender meat. Each individual muscle fibre is surrounded by a sheath, the *sarcolemma* and within this the fibre is divided into *myofibrils* which are surrounded by fluid. The myofibrils are made from two types of protein: *myosin,* which are thicker filaments; and *actin* which are thinner. These proteins are responsible for the contraction of muscle and for *rigor mortis,* which is the contraction after death. A cross-section of a muscle is given in Figure 2.4, showing the fibre bundles. A longitudinal section of the muscle myofibril shows the light and dark striations and the positions of the proteins, actin and myosin, Figure 2.5.

Figure 2.4 Cross-section of muscle

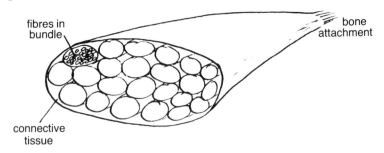

fibres in bundle

bone attachment

connective tissue

Figure 2.5 Longitudinal section of myofibril

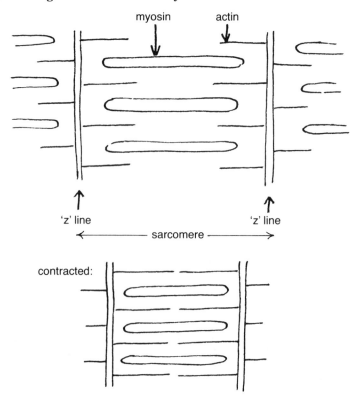

myosin   actin

'z' line   'z' line

sarcomere

contracted:

In Figure 2.5 the ends of the myosin filaments are drawn towards the 'Z' lines with the actin filaments sliding over them. There are six actin rods surrounding each myosin filament. This is repeated throughout the length of the myofibril, and each muscle, and consequently the whole muscle contracts.

Obviously, energy is required for muscles to contract and is released from ATP (adenosine triphosphate) which is kept as an energy source in an inactive complex with magnesium. When an electrical nerve impulse is passed to the muscle to stimulate a contraction a change occurs in the fibre sheath, the sarcolemma. Calcium ions are released as a result of this and these split ATP from the magnesium complex and also stimulate an enzyme, myosin ATP-ase which splits ATP to ADP (adenosine diphosphate) with a release of energy. The magnesium complex of ATP also keeps the proteins, myosin and actin apart, but once it is broken down the two proteins link together at their ends. As energy becomes available the actin is dragged along the myosin filaments and contraction occurs. However, there is still much debate on how this actually occurs.

Once the nerve stimulation of the muscle has stopped the system goes into reverse. Calcium ions are removed from the system so that the magnesium-ATP complex reforms and myosin ATP-ase is inhibited. The proteins again become separated. Respiration processes replenish the ATP at the expense of glycogen in the muscle, but if there has been a lot of muscular activity lactic acid is produced in the muscle with a consequent 'oxygen debt'. Panting, involving taking in a large amount of air, resolves the oxygen debt and metabolises the lactic acid. This is a simplified view of a very complex biochemical process which is still being studied and debated.

2.2.1.2 Conversion of muscle into meat

Although many changes occur in muscle post mortem, the final quality of meat is influenced by a number of factors, pre-slaughter.

Before slaughter animals may lose weight, or become infected with disease from other animals in the abattoir. This loss in weight may cause loss of moisture from the muscles, ultimately leading to tougher or stringy meat, and the reduction of glycogen in the muscle which, as we shall see, can affect meat in a number of ways.

When an animal is killed, or dies naturally, after a period of time the muscles throughout the carcase stiffen in *rigor mortis,* remaining in this condition for some time, after which they soften again. If muscle is cooked while still in rigor it will be tougher and of a darker colour than if it is allowed to pass through rigor before cooking. Meat cooked prior to rigor, however, is always tender, but commercial distribution systems for meat do not generally allow this to happen.

As we saw briefly when discussing the contraction of muscle, ATP is replenished after contraction of the muscle by using up the store of glycogen. When an animal is killed the circulation of blood ceases, but the muscle biochemical processes try to carry on as usual. Glycogen is broken down in the production of ATP as before, but because there is no oxygen available, anaerobic glycolysis occurs and lactic acid is produced. As a result of this lactic acid, the pH of the muscle falls, and enzymes become inhibited. As more lactic acid is produced the production of ATP is inhibited as the enzymes involved are inactivated. The ATP becomes gradually depleted in other reactions and as a result the actin and myosin combine and the muscle contracts, but is unable to relax again – the state of rigor mortis. The ultimate pH in the muscle is important, and is generally about pH 5.6. At a higher pH the meat will not keep well, being subject to microbial attack and having poor colour and water-holding capacity. The more glycogen in the muscle before slaughter the more lactic acid is produced and the ultimate pH is lower.

There are a number of factors which affect glycogen levels in muscle. Only well-fed animals have muscles with the maximum glycogen, and this is readily depleted, particularly in pigs, if the animal is chased, excited or is subject to any stress. Animals kept in pens often fight; this leads to depletion of glycogen when the animals were thought to be resting prior to slaughter. Sugar feeding before death helps to replenish depleted glycogen reserves.

After a few hours rigor mortis disappears and the carcase becomes soft and pliable and the meat is tender, juicier, with more flavour. This period post-rigor is the *ageing* or *conditioning* period of meat and can be an important factor in producing high quality meat. During the ageing process numerous enzyme systems are active that break down large molecules, particularly proteins to release amino acids and lipids to release free fatty acids. The actin filaments are thought to become detached from the Z-line and this leads to increased tenderness in the meat. Connective tissue is not broken down during this ageing period and therefore will still cause meat to be stringy. The release of amino acids, and other nitrogen containing substances and some free fatty acids leads to the development of the typical meat flavour which becomes a main attraction when meat is cooked.

Storage and handling of meat

The processes outlined above can continue leading eventually to deterioration of the meat. Game animals because of their tougher muscle structure need a much longer period of conditioning and are therefore 'hung' for longer periods than other animals and usually at room temperature. Meat must be stored at chill temperatures to slow down the enzyme changes after conditioning and minimise microbial growth. Usually temperatures of 5°C/41°F or less are used, but lower temperatures approaching freezing will greatly extend storage life. Similarly the addition of 10% of carbon dioxide to a chill store will extend storage life but may cause some darkening of the meat. Micro-organisms rapidly contaminate the previously sterile muscles of animals after slaughter. The micro-organisms come from many sources in the abattoir, including personnel, knives, floors, the hide and hooves of the animal, excreta and the air. Some organisms will grow on the surface of the meat and some will penetrate it, ultimately to cause putrefaction. Chilling meat slows down the growth of these organisms but a number can grow at refrigeration temperatures.

The butcher prepares cuts of meat often in a traditional manner and produces a range of meat products. In Figure 2.6 are shown the traditional meat cuts, but these are only a guide as they vary enormously throughout the country. There is a growing tendency to sell meat by its end-use, eg stewing, braising and frying steak and traditional joints are preferred with less fat.

The butcher has to suffer many complaints about his products, the reasons for which are often beyond his control, and often include bad cooking. Toughness is probably the most common complaint. We have discussed some reasons for this previously, but rapid rigor mortis can lead to excessive shrinkage and toughness of the meat and is one of the more common reasons for this complaint. Holding muscle at too high a temperature after slaughter causes this.

In pork a condition known as PSE of muscle, ie pale soft exudate, results if the pH falls too rapidly. This is often a hereditary condition. The proteins denature and lose some of their water-holding capacity causing fluid to exude. In beef, if glycogen is depleted prior to slaughter, the pH remains too high. Too high a pH causes the muscle to have too great a water-holding capacity. As a result of this oxygen cannot penetrate the meat and combine with myoglobin to give bright red oxymyoglobin which gives the attractive bright red colour of meat. The meat, there-fore, remains dark red and tends to be somewhat dry to the palate. The condition produces what is known as 'dark-cutting' beef.

2.2.1.3 Meat products

There are numerous products throughout the world and the market for manufactured meat products such as sausages, pies, canned meat and delicatessen products exceed £1·3 billion per annum. Although the delicacies of many countries are now imported into the UK it is interesting to note that one rarely finds an English style meat-pie on the Continent.

Meat products, usually comminuted in some form or other, are extremely vulnerable to microbial contamination and have been implicated in a large number of food-poisoning outbreaks. The work area must be hygienic and care must be taken to avoid contamination from operators in the factory who might be carriers of some pathogens, eg *Salmonellae.*

Figure 2.6 Traditional meat cuts

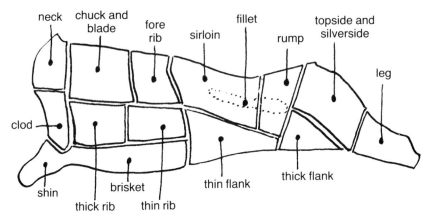

Beef

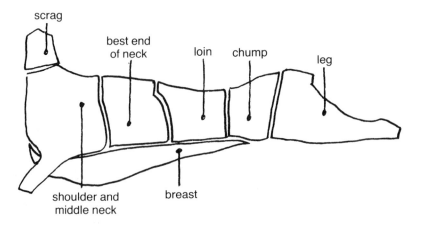

Lamb

Figure 2.6 *continued*

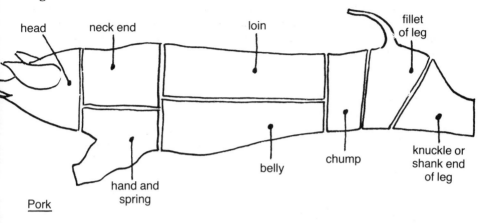

Pork

Curing

A wide range of products, particularly pork derivatives, are cured by the process which started thousands of years ago as salting. Heavy salting of products is not now carried out as the characteristic flavours of cured hams, bacons and smoked meat are preferred.

Bacon

The basis of most current procedures is the 'Wiltshire cure'. The pig carcases are scalded at about 62°C/142°F to remove the hair and any coarse hair is singed. The carcases are then cleaned and split into sides which are cooled and trimmed to remove some bone and unwanted muscle. The temperature is reduced to about 5°C/41°F when curing is started by pumping brine into the sides. This brine varies in composition but is generally about 25–30% salt and 2·5–4% potassium nitrate. The sides are stacked in tanks, perhaps as many as twelve deep, and are covered in brine for about 5 days. Usually the brine is inoculated with salt-tolerant bacteria (halophilic) which are encouraged to grow by the addition of some sugar and by protein leaching from the pork muscle. The bacteria convert the potassium nitrate to nitrite which is broken down to release nitrogen II oxide (nitric oxide) which combines with myoglobin to give the characteristic pink colour of cured meat. The sides of pork are removed from the tanks and matured for about two weeks during which time the typical flavour develops. This 'green' bacon may be consumed or may be smoked. Smoke not only adds flavour but adds phenolic substances which act as preservatives. Liquid smoke flavours are being used in increasing amounts as they are easy and convenient to use. It is possible to charge the bacon slices electrostatically so that when the smoke liquid droplets are sprayed they cling all over the surface ensuring an even 'smoke'. A growing technique is in the preparation of bacon by slice curing in which individual slices are passed for up to 15 minutes through a weaker brine (10% salt and 0·02% sodium nitrite). The overall process is rapid and maturation is complete in a few hours. A good uniform product is produced in a manner which can easily be automated.

Sausage manufacture

Meat included in sausages may come from a number of different parts of the pig's carcase but this meat is always chopped or minced during

Figure 2.7 A sausage production line

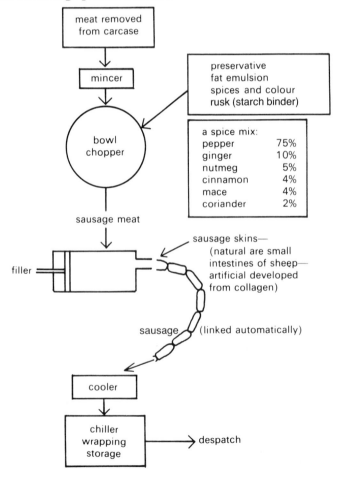

manufacture. This comminution process causes an enormous increase in the surface area available to micro-organisms and the process also distributes these microbes throughout the bulk of the meat, where moist conditions and ample food materials enable rapid growth. The use of chemical preservatives in sausages thus increases what would be a very short shelf-life. Normally the preservative used is sulphur dioxide (not exceeding 450 ppm) but obviously this cannot be added to sausages as a gas but has to be available from a solid sulphite, normally sodium metabisulphite. Fortunately, not only is this preservative active against spoilage organisms which prefer to grow at room temperature, but it is also active against some food poisoning bacteria, particularly *Salmonella*. However, some spoilage organisms are less affected by sulphites and cause sourness to develop in the sausage, thus making them unpalatable before pathogenic organisms can multiply to any extent.

Many types of sausage are available, but in Germany particularly, the number of types is much greater. Some sausages such as bologna and frankfurters are cured then cooked and consequently have a much longer shelf-life even at room temperature. Many products are smoked which lengthens their lives even further.

Canned meats

When meat is sealed into a can (see Section 4.1.2) and heat-processed in order to sterilize the product the manufacturer is faced with the problem that heat penetration is slow and the process takes a long time. This long period of heat-sterilization affects the eating quality of the meat and any reduction in the process would improve the organoleptic properties of the meat, but, of course, would present possible risks of spoilage and food poisoning. A large piece of meat such as a ham is very difficult to process and is normally only pasteurized. The product must, therefore, be stored under refrigeration to prevent spoilage. Nevertheless there is a wide range of canned meats available which include beef, pork and poultry products.

Salami

The production of salami and similar fermented sausages follows to some extent the manufacture of cheese. As with cheese, preservation depends on a fermentation producing lactic acid by bacteria which also produce desirable flavour compounds. Starter cultures are used to bring about rapid lactic fermentation in the meat, which is often held at 30°C/86°F and high humidity to effect what is known as the 'sweating process'. The drop in pH causes a fall in the water-holding capacity of the meat which allows the product to be dried easily. The product is smoked and air dried, and so has a long shelf-life.

2.2.2 Fish

Unlike most meat producing animals, fish have to be caught and for many developing countries they are their only source of animal protein and therefore are very valuable foods. Fish farming is increasingly becoming popular as suitable species become available and economic and disease problems are overcome.

There are many hundreds of fish species used as food and each has its own composition, properties and processing behaviour. The *demersal* fish are bottom feeders and include 'round fish' such as cod and haddock and the 'flat fish' such as plaice and soles. The *pelagic fish* live in the middle and upper layers of the sea and are generally fatty fish such as herrings and mackerel, containing an average 20% lipid. A number of fresh water fish are eaten and besides the more expensive salmon and trout species certain types of carp, perch and pike are popular in some regions. In addition to the fin-fish there are numerous shellfish, such as mussels, cockles and oysters, and a range of crustaceans particularly the shrimps, crabs and lobsters.

2.2.2.1 Fish quality

The most important factor in handling fish is that it is highly perishable and quickly deteriorates. However, there are a number of factors which affect the initial fish quality, particularly the time and place of capture. Some fishing grounds may be low on food, but there is generally a seasonal cycle of feeding with fish moving from one area to another. Spawning occurs at certain times and in certain areas. Fish caught in any of these situations will be 'out of condition', which will be shown in reduced fat content in pelagic species, high water levels and lower protein contents. Fish muscle contains glycogen and the process of rigor mortis also occurs as in meat. However, glycogen reserves may be low in poor quality fish, little lactic acid is produced after death, and the ultimate pH is high giving soft flesh in the fish. Fish obviously cannot be rested before death and in fact some fish 'fight' during capture, in the case of game fish for some time, and also flap or jump about on board after capture. As a result there is significantly less glycogen in fish than in meat and the pH is only about 6·5 compared with 5·6 or less in meat.

Fish rapidly pass into rigor and start to undergo bacterial deterioration immediately afterwards. Most fish are put in ice or frozen to arrest bacterial growth, and this is particularly effective with tropical fish as their inherent bacteria are used to ambient temperatures of 25°C/77°F. Fish flesh contains a nitrogen-containing compound, trimethylamine oxide which is broken down by bacteria into trimethylamine, which has the characteristic smell of bad fish.

Figure 2.8 Production of trimethylamine in fish

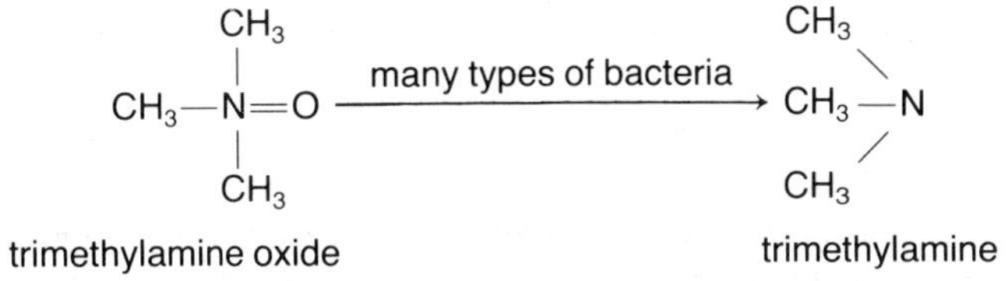

$$CH_3-\underset{\underset{CH_3}{|}}{\overset{\overset{CH_3}{|}}{N}}=O \xrightarrow{\text{many types of bacteria}} CH_3-N\overset{CH_3}{\underset{CH_3}{}}$$

trimethylamine oxide   trimethylamine

Although this substance can be easily detected, it can be determined chemically as an indicator of fish quality.

Ammonia is often produced by bacteria when they attack the protein of fish muscle, again this adds to the smell of bad fish. Any lipid material in fish is usually highly unsaturated, particularly oils in fatty fish, and is therefore susceptible to oxidative rancidity, which again is an indication of poor quality fish.

The typical, but slight, flavour of fresh fish is due in part to inosinic acid. When the fish starts to deteriorate this substance is broken down to hypoxanthine which contributes to the bitter flavour of spoiled fish. The measurement of hypoxanthine levels in fish is an indicator of freshness.

2.2.2.2 Preservation of fish

The lower the temperature the slower the bacterial and enzyme activity in fish and, consequently, the longer the 'shelf-life'. Fish can, therefore, be chilled or frozen.

Chilling (icing)

Ice is an ideal medium for chilling fish and some species may be stored for periods of up to a month. Ice has a relatively large cooling capacity, and for good chilling of the fish the ice must be melting which has the added advantage of washing the fish and keeping it moist. Freshwater fish last longer in ice than salt-water species, and similarly non-fatty fish have a longer shelf-life than fatty fish. Tropical fish, as they are acclimatized to high ambient temperatures, last longer than coldwater species when stored in ice.

In some parts of the world it is not practical to produce and store ice, and an alternative, which is gaining popularity, is the production of chilled sea water (CSW) or refrigerated sea water (RSW). In the former method sea water is chilled by mixing in ice and then the fish are kept fresh in this mixture. To produce RSW a refrigeration plant is necessary to reduce the temperature of seawater to about $-1°C$.

Freezing

Freezing fish, after preparation at sea, and storing at about −30°C/ −22°F can extend the shelf-life of most fish to several months or even a year. Enzymic and bacterial action is almost completely stopped at this temperature and water, required for bacterial growth and enzymic activity, is effectively removed and locked away as ice. The freezing operation may reduce the number of micro-organisms in frozen fish, but on thawing the numbers will more than make up for this decrease. The poorer the quality of the fish, it will be even poorer after freezing and frozen storage. Only the freshest raw fish should be frozen. Often fish are *glazed* after freezing by dipping them into water so that a film of ice forms on the surface of the fish. This glazing prevents *'freezer burn'*, which in fact is dehydration caused by loss of water (ice) from the surface of the fish. Also the technique prevents oxidation of any fat in the fish by acting as a barrier against air.

2.2.2.3 Processing of fish

There are many types of processing methods applied to fish, and this is only a brief review (for further details the reader is referred to the special notes produced by the Torry Fish Research Station at Aberdeen).

Salting

This is a traditional method of processing fish throughout the world, either using salt brines or dry salt. Often the technique is used in conjunction with drying or smoking. Sufficient quantities of salt, through the osmotic effect, prevent the growth of spoilage organisms. Generally a concentration of 6 to 10% of salt in the fish tissues is required.

Fatty fish are slower to salt than non-fatty fish and thicker, fresher fish are similarly slower to salt.

In brine salting the fish are immersed in a salt solution, whereas in dry salting salt is actually rubbed into the surface of the fish. If salt levels are not high enough some bacterial growth may occur producing putrefaction.

Marinades

In addition to salt, acetic acid (vinegar) is used to produce marinated fish. Fatty fish are often preserved in this manner, eg marinated herrings, and have a good shelf-life.

Drying

Sun-drying of fish is an ancient method of preservation which is still practised in many warmer areas. The quality of the product is very variable and, of course, cannot be controlled. Infestation of dried fish by flies is common in many tropical areas. Tunnel driers produce a better quality product but the highest quality of all is produced by freeze-drying (see Section 4.3.1).

Smoking

Smoking fish is another well tried technique of preservation; not only does smoke give flavour to fish but it has a preservation effect due to phenolic compounds. In addition the heat from the fire will dry the product and cook it. The long storage life of some smoked fish is due to the drying and cooking. This type of smoking is called *hot smoking,* but in *cold smoking* the temperature is kept low to avoid cooking and the product must be stored at refrigeration temperatures, eg Finnan haddock.

Canning

Fish for canning must be carefully selected for shape and size and generally must be packed by hand into the cans. The fish must be gutted, cleaned and trimmed, and may be smoked, salted or partially dried prior to canning. In a similar manner to canning meat, the fish is tightly packed into the can and so heat penetration and hence the sterilization process can be very slow. Some fish, eg anchovies, are so heavily salted that no heat-processing is necessary. Fatty fish withstand the heat process much better than white fish, which tend to suffer considerable protein damage.

Fish proteins often contain amino acids with sulphur groups which on heating release hydrogen sulphide. This gas can combine with iron of the can to produce unsightly black stains, and for this reason special lacquered cans must be used. Generally the lacquers contain zinc oxide which produces white zinc sulphide which is not noticed.

2.2.2.4 Crustaceans

This large group of generally expensive sea-foods is characterised by being animals with external skeletons or shells of *chitin*. Most edible crustaceans are ten-legged.

Crabs

As all crabs spoil rapidly, through enzyme activity, when they are dead, it is usual to keep them alive as long as possible. Boiling will inactivate enzymes, particularly if the crabs are boiled in salt water and then they should be chilled. After boiling, crabs or crab meat may be frozen or canned. Generally only the white crab meat is canned.

Prawns and shrimps

These crustaceans are very readily spoiled by their enzyme systems when they are dead; it is therefore customary to boil them immediately in sea-water after catching. They are also found to contain large numbers of micro-organisms which again effect rapid spoilage. An alternative is to remove the heads, which contain most of the digestive enzymes, and chill the tails in ice. Most prawns can be frozen into blocks.

2.2.2.5 Molluscs

Molluscs can be dried, smoked, canned or frozen, but many are eaten fresh. Mussels and oysters are often marinated in vinegar or bottled.

2.2.3 Poultry and eggs

In the muscle of poultry lactic acid is produced after death, and the bird passes through rigor mortis in much the same way as in meat and fish. In chicken, particularly, the breast muscles are little used and so they only need a small amount of haemoglobin and myoglobin. In consequence the breast muscle is white whereas the leg muscle produces dark meat. The increase of poultry sales has been significant during the last decade and more recently the sales of poultry products, particularly from turkey, have made significant increases.

Nutritionally chicken and beef muscle are similar, but as a source of protein chicken protein is produced more quickly and more cheaply.

2.2.3.1 Processing

The poultry-processing industry is one which has become rapidly mechanised with a demand for large numbers of young, uniform chickens. Automated processes are used for rapid killing, dressing and chilling of the carcases. Rapid production and quick freezing minimise problems of infection by micro-organisms.

Birds are generally not fed for several hours and are initially electrically stunned, followed by cutting the head to sever the main blood vessels. Blood is drained away and the birds are scalded with boiling water. The feathers are plucked and this can be mechanised but requires subsequent visual inspection. Feet, heads and viscera are removed. The carcases are often cooled in iced water to reduce enzyme action and colour changes. The chickens are then dried, trussed and packed into polythene bags, prior to freezing in a blast freezer (see Section 4.2.2).

2.2.3.2 Eggs

The hen's egg performs two important purposes, that of serving as a means of reproduction for the species, and secondly, providing man with a nutritious, convenient and cheap food. The egg is equipped with means of defending the embryo chick against micro-organisms during its development, but some of the systems may affect the nutritional aspects of the egg for man.

Structure of the egg

The outer shell of an egg is composed mainly of calcium carbonate, lined with membranes and having a number of pores for gas exchange. These pores are covered by a wax-like layer known as the *cuticle*. This cuticle protects, in part, against microbial invasion and controls water-loss. The structure of an egg is given in Figure 2.9.

Figure 2.9 Structure of the hen's egg

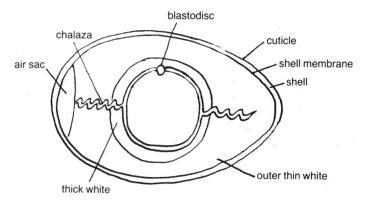

The egg white is divided into a thick layer around the yolk and a thinner layer next to the shell. This thick layer anchors the yolk, together with the chalazae, in the middle of the egg where it is less vulnerable to microbial attack. Egg whites foam readily and are therefore useful for making souffles and meringues. However, in the egg the white has an important function as it possesses special antibacterial properties, which under normal circumstances prevent the growth and multiplication of micro-organisms which have penetrated the shell. The enzyme, *lysozyme,* actually destroys bacteria by lysis (splitting) of their cell walls. The *ovomucoid* is a protein in the white which inhibits the enzyme *trypsin* and the protein *avidin* binds up the vitamin biotin. *Conalbumin* acts as a sequestrant and binds up iron and copper. All these factors are anti-microbial, but of course will affect the egg's nutritional properties. Fortunately, cooking an egg destroys the properties of ovomucoid and avidin. The pH of eggs is unusual in being alkaline, often as high as 9·0, again an anti-microbial property. The yolk is rich in nutrients and is the most vulnerable part of the egg to micro-organisms. The yolk has good emulsifying properties, and so is used in the preparation of mayonnaise, salad dressings and cakes.

Quality of eggs

Biological changes gradually take place inside the egg and unless these can be controlled, loss in quality results. The thick white of the egg generally becomes thinner and membranes around the yolk weaken. Loss of water during storage causes an enlarged air sac, which can be reduced by humidification of the store.

Candling of eggs is a method commonly used for checking the quality of eggs, although it is no longer so widely employed. The egg is illuminated with a light so the following flaws can be seen: the yolk position and size; the size of air sac; and the presence of blood or 'meat' spots.

The yolk index is another method of determining quality. The height of the yolk is measured with a special tripod micrometer, and its width is measured with calipers. The index is calculated thus:

$$\frac{\text{height of yolk}}{\text{width of yolk}} = \text{yolk index}$$

A fresh egg usually has an index of about 0·45 and this gradually falls on ageing as freshness decreases.

In 1988 the problem of Salmonella (*S enteritidis*) in egg yolks was first recognised, this has now been brought under control.

Egg processing

Frozen whole egg is produced in large quantities for use in food manufacture. Eggs are washed and any bacteria on the surface are destroyed by chlorine in the wash water. The eggs are broken mechanically and mixed to an homogeneous product which is pasteurized at 63°C/145°F for 60 seconds. Any shell pieces are filtered out and the liquid egg is filled into cans and frozen in a blast freezer.

Dried egg is an alternative to frozen egg and is usually made by spray-drying, although freeze-drying produces an excellent but expensive product. Eggs are prepared as above and spray-dried in a drier which can be used for milk powder.

Review

1. Meat
   – muscle, associated connective tissue, and adjoining fat
   – number of biochemical changes necessary to form meat
   – good supply of most nutrients
   – connective tissue makes it indigestible

2. Muscle structure

- muscle in meat – striated muscle
- myofibrils make up muscle fibres, held together by sheath, the sarcolemma
- two proteins in myofibrils – myosin and actin
- these two proteins responsible for muscle contraction
- rigor mortis, contraction after death
- ends of myosin drawn towards the 'Z' line with actin filaments sliding over them

3. Conversion of muscle to meat

- high glycogen level important as leads to high lactic acid and final pH of 5·6
- higher pH leads to poor colour
- animals must be rested prior to slaughter to ensure high glycogen levels in muscles
- conditioning or ageing after slaughter
- enzymes breakdown large molecules to smaller molecules, give flavour and tenderness to meat

4. Storage and handling of meat

- chill to 5°C or less to slow down enzymic changes and microbial growth
- rapid rigor mortis can cause excessive shrinkage and toughness

5. Meat products

- usually comminuted – greater risk of food poisoning
- 'Wiltshire cure' base of most bacon and ham curing
- smoking adds flavour and preserves by adding phenolic compounds
- sausages require preservative usually sodium metabisulphite
- canned meats require long processing times as the cans are tightly packed
- 'salami' manufactured in similar manner to cheese

6. Fish

- demersal fish – bottom feeders include cod, haddock and flat fish
- pelagic – middle and surface feeders, generally fatty fish including herrings and mackerel
- fish can be 'out of condition' when caught during certain seasons and during spawning
- glycogen reserves are low and quickly used up causing rapid rigor mortis
- final pH not as low as in meat
- trimethylamine oxide broken down to give trimethylamine – smell of bad fish
- ammonia also produced in bad fish
- inosinic acid, partly responsible for fish flavour, broken down to hypoxanthine – contributes to bitter flavour of spoiled fish

7. Preservation of fish

- ice ideal medium for cooling
- freshwater fish last longer in ice than salt-water
- non-fatty fish last longer than fatty fish
- freezing at sea and storage at −30°C extends storage life of fish for several months

– preservation methods include:
salting, marinating, drying, smoking and canning
– crustaceans undergo rapid enzymic spoilage and microbial attack

8. Poultry and eggs

– poultry similar to meat but white breast muscle has little respiratory pigment
– eggs perform function of reproduction and food for man
– egg white contains anti-microbial systems in its proteins:
   (1)   lysozyme – destroys bacteria
   (2)   ovomucoid – inhibits enzyme trypsin
   (3)   avidin – binds vitamin, biotin
   (4)   conalbumin – binds iron
– egg yolk good emulsifying properties

Practical exercises: *Meat, fish and eggs*

1. Microscope examination of meat and fish muscle

Place a small amount of tissue on a slide, add a drop of water and attempt to spread the tissue into a thin section. Examine the fibres under high power of the microscope.

2. Cooking losses in meat products

Losses are caused by moisture and fat separating from the product, producing noticeable shrinkage. Try the following experiment on sausages, meat burgers and minced meat.

Weigh the product. Weigh 60 g of lard and add to a frying pan and heat. Fry the product for 10 minutes. Reweigh the drained product and weigh the fat from the pan *(when cooled)*. Calculate the weight loss of the product, fat loss and moisture loss.

3. Fish filleting

Evaluation of the degree of wastage associated with filleting different species. Prepare two fillets from a weighed fish. Weigh the fillets and calculate the percentage yield of fillets. (The fillets may then be frozen or used in other work).

4. Eggs – physical characteristics

Note nature of egg shell. Carefully break an egg on to a plate. Measure the yolk height and width, and calculate the yolk index. Test the egg parts with universal pH papers and note the pH. Try to observe each part of the egg structure.

5. Emulsifying property of egg yolk

Mix a drop of oil with some water, and shake. Note the speed of separation. Repeat using washing-up liquid, and then with some egg yolk.

Note emulsification produced.

2.3 Fruit and vegetables

The botanical term fruit refers to the organ which carries the seeds. However some fruits are used as vegetables, for example tomatoes. Vegetables are from all other edible parts of a plant, such as the leaves, stem, roots and any storage organs. It is simpler to divide the two as follows: fruit have fragrant flavours and are usually sweet (or have sugar added); vegetables are soft plant products eaten with animal products and are not sweet but usually salted.

Many special varieties of fruit and vegetables have been developed by plant breeders for individual food processes. Carrots, for example for canning, must be small, of even colour throughout and straight-sided. Darker peas have been developed for freezing, but paler types are better for canning. The crop must ripen at one and the same time to permit one destructive harvest, whereas it is beneficial for garden crops to ripen a little at a time so that they can be used in a fresh condition.

2.3.1 The composition of fruits and vegetables

Being principally water, most fruit and vegetables provide to the average diet only minerals and some vitamins, and perhaps some roughage. However, these additions, particularly of vitamins such as ascorbic acid, can be of immense value in some diets. Some products, however, are good sources of carbohydrate, eg banana. In some parts of the world some crops form a significant part of the diet and in world tonnage potato, cassava and banana are the most consumed. Some 300,000,000 people are thought to live mainly on cassava in the equatorial belt of the world.

Plant tissues assume a characteristic maximum water content and at this level the tissue is said to be *turgid* or in a complete state of *turgor*.

A typical plant cell is somewhat like a balloon, however, it is not blown up with air but with water. The internal pressure in the cell, called *turgor pressure* can be as high as nine times atmospheric pressure. In a normal cell in full turgor this pressure is equalized by the elasticity of the cell wall, Figure 2.10.

Once the supply of water is reduced or cut off to the cell, water is gradually lost from the plant by *transpiration* and the turgor pressure cannot be maintained. The elasticity of the cell wall now exceeds the outward turgor pressure and the cell starts to collapse. This is repeated throughout the plant which 'wilts' like a newly planted cabbage seedling on a hot day. Harvested leafy vegetables, such as lettuce remain turgid for only short periods before becoming limp. As plant cells become older a complex substance *lignin* is deposited on the cell wall, thus making it

Figure 2.10 Typical plant cell in full turgor

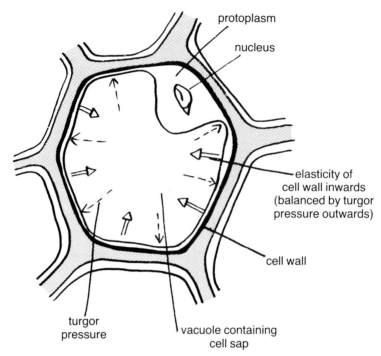

tougher. However, lignified tissues are old and poor to eat, whereas most vegetables should be eaten when small and tender.

Water enters the plant cell by the process of *osmosis* as the cell contains a concentrated solution of sugars and acids in the form of *cell sap*. The cell membranes maintain the cell in fully turgid state but also allow the exchange of other substances.

The *texture* of fruit and vegetables is often an index of quality and it depends not only on the turgor of the cells but also on the presence of supporting tissues and the cohesiveness of the cells. About one third of the measured texture (firmness) of a small fresh carrot is due to turgor pressure.

The *colour* of fruit and vegetables is an important attractive feature, and most pigments occur in *plastids* which are specialized bodies lying in the protoplasm of the cell. Chlorophylls occur in the chloroplasts which are bodies containing the green pigment. The cartenoids are also present in these chloroplasts but until the fruit ripens are masked by the chlorophylls. Many of the bright reds, violets and blue colours are due to the anthocyanin group of colours (see Section 1.7.1.3).

Simple sugars are present in most fruit and vegetables, particularly sucrose, fructose and glucose. The latter two, reducing sugars, usually predominate but in some fruits sucrose is most dominant, for example in onion, carrot, pea, banana and melon.

The *cell wall* of fruit and vegetable cells is built up from a number of carbohydrates, cellulose being the main one, supported by pectins and hemicelluloses. Cells are held together by pectins and hemicelluloses which occur in the *middle lamella*. Changes in these occur in ripening so the cells part easily, leading to a softer texture. Starch is the second most common carbohydrate after cellulose and is the main food material in most vegetables, particularly in storage organs such as tubers, for example potatoes and Jerusalem artichokes.

There is very little *lipid* in fruit and vegetables, but some occurs as 'cutin' which controls water loss from fruits such as apples. The *protein* content of fruit and vegetables varies enormously, and unfortunately some staple foods such as cassava are very low in protein and lack many of the essential amino acids. Legumes, ie peas and beans, are richer in proteins than most plant products, containing often about 8% protein. Much of the protein, however, is in the form of enzymes which control the composition and ripening after harvesting in many fruits and vegetables.

Fruit and vegetables are normally acidic in nature, having a pH between 2 and 4. Malic and citric acids are the most common acids, sometimes one predominates, sometimes the other. Avocados are deficient in both and grapes when ripe, have tartaric acid. Citrus fruits, blackcurrants, raspberries, pears, potato, legumes and tomato have citric acid as their main acid. Malic acid dominates in apples, apricots, cherries, plums, bananas, lettuce, onion and carrot. Many other acids have been reported in smaller quantities, and many have important roles in metabolism (refer to Kreb's cycle). The most important acid, which is also a vitamin, is of course ascorbic acid. The outside layers of most fruits contain more than the inside, similarly fruits grown in the sun are richer than those in the shade. The West Indian Cherry (Acerola) contains 1500 mg/100 g of ascorbic acid which compares with most oranges at 40–60 mg/100 g. As a rough guide, highly coloured fruit are richer in vitamin C than less coloured fruit.

The flavour compounds in fruits are largely oxygenated compounds such as esters, alcohols, acids, aldehydes and ketones, many being derivatives of the monoterpenes (see Section 1.8.2.1). Vegetables have a limited range of volatile odoriferous compounds, esters are generally lacking, but individual acids, alcohols, aldehydes and ketones may be found. Many sulphur-containing compounds may occur in the cabbage family and onions (see Section 1.8.2.2).

Some fruits and vegetables unfortunately contain small amounts of complex substances which are *poison precursors*. Cassava contains the substance *linamarin* which is a glucose derivative containing a cyanide group. In certain traditional methods of preparing cassava an enzyme breaks down this substance to release free hydrocyanic acid. Gradual cyanide poisoning can occur as a result of this and in its mildest form can cause goitre and cretinism in people already on poor diets. Potatoes contain the glycoside solanine which can cause mild stomach upsets but in green potatoes levels of solanine can reach more dangerous levels.

2.3.2 Post-harvest changes in fruit and vegetables

When detached from the tree, or dug from the ground, a fruit or vegetable can continue to live and respire for a period of time, which, for some products, may be a number of months. This period of existence obviously uses up storage material within the product and gradually its quality falls until finally it undergoes senescence and decay. The length of storage of a plant product depends on: its chemical composition; resistance to microbial attack; external temperature; and presence of various gases in the storage atmosphere.

Respiration is the major process of interest in post-harvest fruit and vegetables. The rate of respiration, which involves the oxidation of energy-rich organic compounds to form simpler compounds and yield energy, is indicative of the rapidity with which compositional changes in fruit and vegetables are taking place.

Figure 2.11 The climacteric in fruits

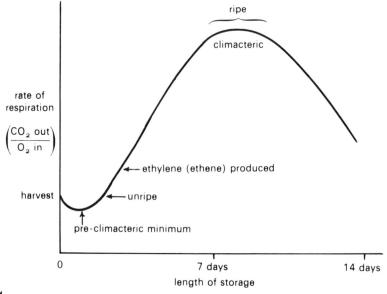

Fruit and vegetables can be divided into two groups according to their respiratory behaviour. *Climacteric* fruits are generally 'fleshy' fruits which show a rapid rise in the rate of respiration after harvesting which leads to ripening and then senescence of the fruit. *Non-climacteric* fruits, and most vegetables (except tomato) show no rapid increase in respiration and changes in ripening and maturation are gradual. The characteristic climacteric pattern is shown in Figure 2.11, but, with the exception of the avocado, few fruits follow exactly this pattern.

Climacteric fruits include banana, mango, pear and avocado. The fruit normally attains the stage of ripeness best for eating at the climacteric or some time after the peak. Ripeness is a somewhat variable term, as ripeness for eating is overripe, for example, for jam-making where more pectin is required, but underripe for juice-extraction where a soft, juicy fruit is required.

Non-climacteric fruit include the citrus group, pineapple, fig and grape, and also most vegetables do not show a climacteric. However, some fruits show a very weak climacteric, for example, most species of apples show a slow and shallow peak. Fruits which do not show the climacteric keep well for long periods under normal conditions, whereas climacteric fruit quickly ripen and deteriorate presenting many problems in storage.

It has been known for some years that the gas ethylene (ethene) acts as a plant hormone and stimulates the ripening process. Ethylene is produced in the fruit at the beginning of the climacteric peak and is thought to be produced from the amino acid methionine, as shown in Figure 2.12. Ethylene can be applied externally to cause ripening. Unripe fruit rapidly ripen when stored with already ripe fruit.

Figure 2.12 Production of ethylene (ethene)

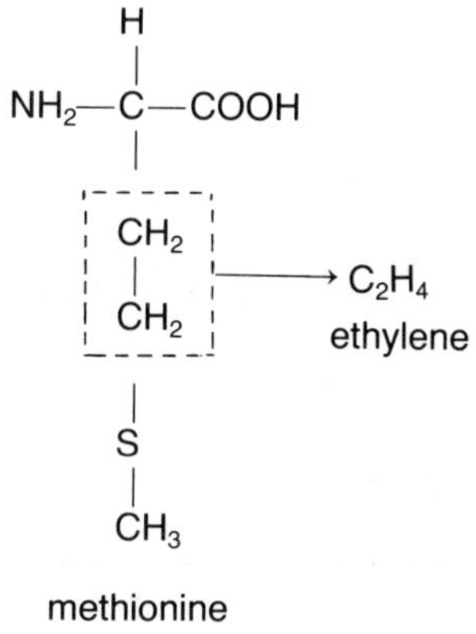

methionine

As ripening proceeds enzyme systems become active, resulting in a number of changes. The most important changes are in the carbohydrates contained in the fruit. Sugars increase at the expense of polysaccharides such as starch and cell wall polysaccharides are broken down which leads to a softer texture in the fruit. The pectic substances are progressively broken down as ripening proceeds, as shown in Figure 2.13, which eventually leads to a very soft product.

Figure 2.13   Pectin breakdown during ripening

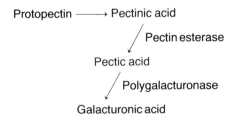

Acid levels fall during ripening in some fruits, but are masked by the increase in sweetness due to sugars being produced. The most obvious changes in fruits during ripening are in their pigments, as chlorophyll is broken down to reveal other colours such as the yellow/reds of the carotenoids. Some pigments are synthesised during ripening, for example the anthocyanins and lycopene.

Changes are gradual in most vegetables and non-climacteric fruit. Changes occurring depend on which part of the plant the vegetable constitutes, as shoots, for example, continue to grow and lengthen. In vegetables with pods, such as beans and peas, protein is broken down in the pods and the resulting amino acids are translocated to the seeds and resynthesised into protein. Starch is synthesised from sugars during the post-harvest period of vegetables, which is the converse of what occurs in fruits. However, there appears to be a competition between the synthesis

Figure 2.14   Respiration and starch synthesis competition in vegetables

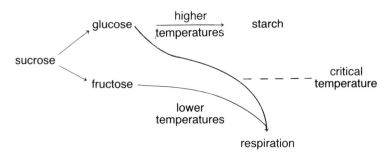

of starch and the use of sugar for respiration and temperature appears to control this.

The *critical temperature* is a characteristic of each vegetable, and in potato it varies from about 1·5–4·5°C. In tropical products the critical temperature is much higher, usually around 13–16°C. Above the critical temperature starch is synthesised, but below the temperature starch is broken down, sugars accumulate and respiration increases. This is clearly shown in winter, when potatoes are stored at too low a temperature. They contain too much sugar, ie reducing sugars glucose and fructose, and produce very dark chips and crisps on frying.

2.3.3 Storage of fruit and vegetables

Until they are consumed or processed, fruit and vegetables are living, respiring biological systems. Respiration, and consequently ripening may be inhibited by: controlling temperature during storage; lowering the available oxygen in the store; increasing carbon dioxide levels; lowering atmospheric pressure; or using special coatings.

2.3.3.1 Temperature

Respiration slows down as the temperature of the environment decreases and thus ripening and senescence are delayed. However, chilling some fruit, particularly those of tropical origin, may cause considerable damage below a certain temperature. Below about 13°C/55°F some enzyme systems are inhibited in tropical fruit while others continue, causing the accumulation of toxic intermediate products. This can cause discolouration, flavour and texture change in the fruit leading to rots and bacterial decay. Try storing a banana in a refrigerator!

2.3.3.2 Controlled atmosphere storage

Controlling the atmosphere in a store can slow respiration and delay ripening. The oxygen level can be lowered or the carbon dioxide level can be increased. This is called *controlled atmosphere* (CA) or *modified atmosphere* (MA)* storage. It may be used in conjunction with temperature control. However, physiological disorders can readily occur if the gas levels are incorrect, as too high a CO_2 or too low an oxygen level can cause *anaerobic respiration* leading to the accumulation of damaging toxic compounds in the fruit. A simple way to make a CA store is just to allow the fruit to respire for a time to lower the oxygen and increase the CO_2 levels. This is generally unsatisfactory as respiration has started and will try to continue at a higher rate leading to spoilage. Correctly blended gases from cylinders are usually fed into a chamber from which most of the air has been removed. Together with

* This term is now being used for modified atmosphere within packaging films.

from which most of the air has been removed. Together with temperature control this is a very good system for storing many fruits and vegetables. Excess CO_2 is often removed by 'scrubbing' with sodium hydroxide which absorbs the gas.

2.3.3.3 Hypobaric (low pressure) storage

This is a type of CA storage in which the atmospheric pressure in the store is reduced, so lowering the oxygen level and causing ethylene to diffuse out of the fruit, thus delaying ripening. Although the method has produced some good results, it is very expensive and difficult to perfect.

2.3.3.4 Special coatings

For a number of years some fruit have been waxed to prevent dehydration and to retard ripening. One of the most significant developments in food technology in recent years has been the development of an edible coating which can retard the ripening of most fruits.

A mixture of lipids (sucrose esters of some fatty acids) mixed with a polysaccharide is now being marketed under the trade mark *Pro-long*. The powdered mixture is dispersed and dissolved in water and into this the fruit and vegetables are dipped and allowed to dry. The result is the formation of a semi-permeable gas barrier which allows oxygen to diffuse into the fruit but retains some of the carbon dioxide produced during respiration. This mixture of internal gases reduces the metabolic rate of the product. It has a number of other advantages, such as reducing water-loss and controlling some spoilage diseases. The treated produce can be stored at higher temperatures and therefore the method offers considerable savings on any other system at present in use. Ideally, crops should be treated as soon as possible after harvesting.

The potential of extending the storage life of many crops by up to three times by this method offers many new possibilities for the marketing of unusual tropical fruits. Developing countries can make use of a simple technique like this in order to develop an export trade of fruit and vegetables which are cheap to produce and often under-exploited.

2.3.4 Fruit and vegetable processing

Fruit and vegetables need to be peeled before they undergo further processing and preservation by canning, drying, freezing or in preserves. In the kitchen it is simple to peel by hand, but obviously this would be labour-intensive and expensive in a large factory. Boiling water and steam can be used to 'scald' produce in about ½–3 minutes, in order to loosen the peel which can then be removed by rotating brushes and cold water sprays.

'Lye peeling' involves the use of a solution of caustic soda (sodium hydroxide) which disolves the walls of the cells making up the peel, causing the peel to disintegrate. This common method will remove the peel and areas of bruised tissue in an economical manner. The lye can be neutralised with a solution of citric acid after peeling.

It is possible to freeze-peel by spraying liquid nitrogen on to the fruit, which on thawing can be easily brushed to remove the damaged peel. Acid solutions, particularly hydrochloric (0·1%), citric (0·1%) and tartaric (0·1%) have also been used for peeling a number of products and have the advantage of inhibiting any enzymic browning which might occur.

Dehydration

Many dried fruits are produced by the traditional sun-drying process in hotter countries. Obviously there is little or no control over this method and quality of the product is variable. Tunnel driers (see Section 4.3.1) have been used extensively for drying fruit and vegetables, which are usually in a cube form, ie diced. The highest quality dried products are produced by freeze-drying and particularly by accelerated freeze-drying (AFD) (see Section 4.3.1). However, the relatively high cost of the process necessitates a higher price for the product, which is not usually justified for cheaper fruits and vegetables.

Juice extraction

Juices can be squeezed from many different fruit and vegetables by a number of methods. It is often beneficial to macerate the product prior to extraction or to treat it with enzymes, particularly pectolytic enzymes, to increase the yield of the extracted juice. There are two main groups of juices, low viscosity and high viscosity juices.

Low viscosity juices, such as apple juice, are clear without any suspended matter. The apples are macerated to form a pulp which is pressed in a rack-and-cloth press to release the juice. Pectolytic enzymes and sometimes amylases are added to clear the juice, and any suspended particles are removed by filtration. The juice is then pasteurized prior to bottling or canning. Centrifuges are sometimes used in addition to the press to increase the yield.

Tomato juice is a typical *high viscosity juice* and the higher the viscosity the better the quality. The consistency of the juice is controlled by the method of manufacture and is not solely related to the solids content of the juice. In the *cold-break* method raw tomatoes are macerated at room temperature, then the seeds and skin are filtered from the juice. This process yields a juice of lower viscosity as the pectins are degraded by pectolytic enzymes; however the juice has excellent flavour and colour.

In the *hot-break* method, most favoured by industry, tomatoes are macerated and heated to 85°C/185°F to inactivate the pectolytic enzymes. The viscosity of juice is higher but the flavour and colour are not as good as in the cold-break method. Often hydrochloric acid is added during maceration, as this improves consistency, and is then neutralised after juice extraction with sodium hydroxide solution, which, with the acid, forms sodium chloride giving a degree of saltiness to the product.

Pickling

Pickles are usually made from cucumbers, onions, cauliflower and cabbage, but virtually any product can be used, and the practice of pickling dates back many thousands of years. The product in a pickle is preserved and flavoured by a solution of salt and edible acid, usually vinegar. The acid is either added or produced by fermentation, as nearly all vegetables can be fermented by lactic acid bacteria to yield a sufficient level of lactic acid. Preservation of vegetables by fermentation depends on: the reduction of the activity of the natural enzymes of the product; inhibition of oxidative chemical changes; and inhibition of the growth of spoilage organisms. Pickles can be prepared directly from vegetables, without fermentation, by adding salt and vinegar. Some products are allowed to ferment in a weak brine solution, for example, the expensive 'dill' pickles. Other products are fermented in a high-salt brine and then later converted into mixed pickle, eg 'salt-stock' pickles.

A traditional method is to fill a wooden vat about ⅓ full of a 10% brine into which the cucumbers or other vegetables are filled. Dry salt is added to try to keep the salt solution at 10%. Brine is again added to cover the cucumbers completely so that fermentation can occur in the absence of air. As the bacteria multiply the brine becomes cloudy. Lactic acid is produced, and together with the salt, preserves the product.

Preserves

Many fruits are made into preserves which include jellies*, jams, marmalades, conserves and candied fruit. As gel formation is an essential part of a preserve, underripe fruit, rich in pectin, is the ideal material to use but lacking this, pectin may be added. Fruit pulp is often frozen or canned and stored to be made into jam out of season. Often sulphur dioxide is added to prevent microbial growth and discolouration, but is boiled off during jam manufacture (see Section 1.2.2.2).

In the traditional *jam-making* process sugar is added to an equal weight of fruit, but in reality the amount of sugar used depends on: the acidity of the fruit; the sugar content of the fruit; ripeness of the fruit; and the type of product being made. The mixture is boiled, generally under reduced

*NB Table jelly is made from gelatine **not** pectin.

pressure, to reach the required total soluble solids content of about 68%.

Marmalades contain fruit pulp and peel, and being made from underripe fruit are particularly rich in pectin and acid. Citrus fruits are commonly used but pineapple, ginger, pear and grape marmalades are gaining popularity.

Conserves are generally similar to jams, but as they contain much more fruit they are expensive. Often chopped nuts, such as walnuts, are added to give texture and flavour to the product.

Crystallized or candied fruit is a traditional product, particularly in the Middle East. Strong sugar solutions are added to prepared slices of fruit which become dehydrated by the osmotic effect of the solutions and are thus preserved. Glacé cherries, citrus peels and mixed cake decorations are produced in this manner. In the manufacture of glacé cherries the cherries, destoned, are heated to 60°C/140°F for up to 20 minutes and a small amount of calcium chloride is added to firm the tissues. The cherries are then added to boiling syrup and may stand in the syrup for some time. The prepared cherries are very sensitive to high humidity and must be stored at below 50% Relative Humidity and 10°C/50°F.

Review

1. Composition of fruits and vegetables

- mainly water, contribute minerals and vitamins, some roughage. Some products good carbohydrate sources, eg banana
- turgid cell – at maximum water content
- turgor pressure in cell balanced by elasticity of cell wall
- transpiration of harvested produce causes loss of turgor and wilting
- colour pigments occur in plastids
- cell wall made up of cellulose, hemicellulose and pectin
- malic and citric acids most common
- highly coloured and tropical fruits richer in Vitamin C
- some products contain poison precursors eg linamarin in cassava can break down to release hydrocyanic acid

2. Post-harvest changes

- length of storage depends on:
  - (1) chemical composition
  - (2) resistance to microbial attack
  - (3) external temperature
  - (4) presence of gases in the store

– respiration major process of interest
– climacteric fruits show rapid rise in rate of respiration after harvesting
– non-climacteric fruits and most vegetables show no rapid increase in respiration, but changes in composition are gradual
– ethylene (ethene) stimulates ripening process
– pectin broken down progressively
protopectin → pectinic acid → pectic acid → galacturonic acid
– chlorophyll broken down to reveal other pigments eg carotenes
– or others, eg lycopene are synthesised during ripening
– in vegetables, such as peas and beans, starch is synthesised from sugar
– at lower temperatures starch is not synthesised and sugars accumulate, respiration increases

3. Storage

– respiration and ripening controlled by:
(1)   reducing the temperature
(2)   increasing CO_2 levels
(3)   decreasing O_2 levels
(4)   lowering atmospheric pressure (hypobaric storage)
(5)   use of special surface coatings such as 'Pro-long'

4. Processing

– peeling, essential prior to further processing, can be by:
(1)   hand peeling
(2)   steam or hot water
(3)   lye peeling using sodium hydroxide solution
(4)   freeze peeling using liquid nitrogen
(5)   acid peeling
– dehydration – traditional products made by sun-drying
– many products produced using tunnel driers
– best products freeze-dried (AFD)
– juice extraction
– low viscosity juices, eg apple, are clear and need the action of pectolytic enzymes
– high viscosity juices require pectin to maintain consistency, therefore, pectolytic enzymes should be inactivated, eg in hot-break method for tomato juice
– pickles – rely on salt and acid for preservation
– acid may be added, eg vinegar, or produced by fermentation,
eg lactic acid
– preserves
– use underripe fruit or add pectin
– conserves rich in fruit with added nuts
– candied fruit, slices of fruit dehydrated by strong solutions

Practical exercises: *Fruit and vegetables*

1. Microscopic examination

Prepare the sections of a number of products, using a scalpel.
Examine cellular structure. Stain some samples with a drop of iodine and examine for starch.

2. Storage of fruit

Take samples of two different fruits, eg apples and bananas.
Store for several weeks: at room temperature; in a refrigerator; near a boiler; in an air-tight tin. Observe the overall quality of the product every week and note changes taking place. Cut open some of the fruit and note internal changes.

3. Peeling of vegetables

Compare the efficiency of peeling methods by weighing a sample of vegetables (potato, carrot or small turnips) before and after peeling.

Methods

   (a) Hand peeling

   (b) Lye peeling – immerse vegetables in a boiling solution of 5% sodium hydroxide solution (CARE!) for a timed period. Remove and drain, wash off skin under a running tap and weigh product.

   (c) Steam peeling – place vegetables in a pressure cooker and steam for 2, 4, 6 and 8 minutes. Wash off peel under the tap and weigh vegetables.

   (d) Abrasive peeling – if available use a peeling attachment for a domestic mixer or processor.

Peeling must be efficiently carried out, but with the minimum weighed loss. Which method meets these criteria?

2.4 Cereals and baked products

Cereal grains are one of the most important sources of carbohydrate and hence energy in the diet. Cereals belong to the grass family, *Gramineae,* and there are many thousands of varieties which have been cultivated since the beginning of civilization. The grains may be consumed whole or processed by many methods into flour, starch, bran, sugar syrups and some oil. Rice alone feeds over half the world's population, particularly in the East where 93% of all rice is produced. The area of the world devoted to wheat production is double that of rice, but rice yields over double the grain produced from wheat.

2.4.1 Composition of cereals

Most of the cereals have similar chemical compositions as they belong to the same family of plants. However, composition depends upon variety, weather and growing conditions. The moisture content of cereals is always higher after harvesting, but if properly handled the grain will attain a lower figure of about 10–14%, at which level germination of the grain is inhibited and moulds and bacterial rots do not usually occur.

The composition of cereals, in general terms, is given in Table 2.7.

Table 2.7   General composition of cereals

Constituent	%	Example showing highest figure
Carbohydrate	58–72	Maize
Protein	7–13	Wheat
Moisture	10–14	Barley
Fat	1– 7	Oats
Fibre	2–10	Oats
Vitamins:		
Thiamin	3–90 µg/g	Oats
Riboflavin	1·1–1·6 µg/g	Rye
Niacin	15–100 µg/g	Barley
Pantothenic acid	4·4–17 µg/g	Rice

2.4.2 Structure of a cereal grain

All cereal grains share three main features. The first feature is an embryo or germ (hence wheatgerm) from which the root and shoot sprout. Secondly, the endosperm is a store of starch which is broken

Figure 2.15 Structure of a wheat grain

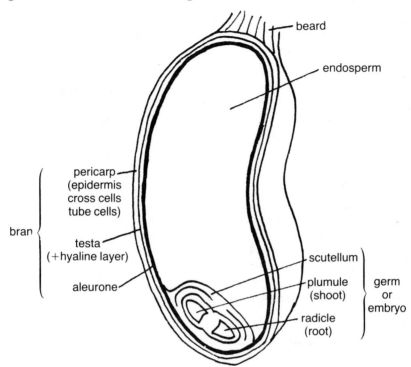

down to glucose to supply energy for the growing embryo. The third feature is known collectively as the bran, which is just several layers which cover and protect the grain. The wheat grain shows these features well and in Figure 2.15 the various parts of the grain are labelled.

The distribution of nutrients is obviously uneven throughout the grain, as the starch is concentrated in the endosperm and many of the B complex vitamins are in the outer bran layers, which is clearly illustrated by thiamin distribution. About 60% of thiamin is in the bran, 24% in the endosperm and 16% in the germ.

The germs of many cereal grains contain vitamin E (tocopherols) but this can readily be lost by oxidation.

2.4.3 Wheat milling

Although rye has some bread-making qualities, most bread is made from wheat. The wheat grain has to be milled and separated into a number of suitable fractions to make flour which can be used for baking.

When we buy flour from the supermarket we expect it to be of a consistent quality and able to produce bread, or cakes or biscuits as required. To produce flours of the required different characteristics a number of different wheat strains are blended in the mixture fed into the mill (the grist).

'Strong' wheats have a higher protein content and produce bread of good loaf volume and good texture. 'Weak' wheats are lower in protein and are more suitable for cakes and biscuits. American wheats are particularly strong wheats whereas European wheats are weak. However, there are new European wheats, for example the English variety 'Avalon' which are stronger than former varieties and are good for bread-making. It is possible to extract protein from strong wheats and add to weak wheats to improve their bread-making performance.

The terms 'hardness' and 'softness' of wheats are sometimes encountered and these refer to milling characteristics of the grains. Hardness refers to the ease with which the endosperm disintegrates during milling. In hard wheats the endosperm separates more easily and intact from the bran, but in the soft wheats the endosperm breaks down easily and does not separate.

The milling process can be divided into preliminary cleaning and conditioning, and secondly, size reduction of the grain and separation from the bran.

Preliminary cleaning and conditioning

A number of different specialized machines are used to remove all contaminants such as stones, string, metal, straw and other seeds. Sieving and blowing air through the grain will remove some contaminants. A disc separator uses discs with indents into which the wheat grains fall while other grains, such as oats and barley, pass through the machine. The discs rotate and then discharge the caught wheat grains, which then pass to magnets to remove metal particles. The grains are washed in a moving stream of water, and then surplus water is removed in a type of centrifuge, the 'whizzer'. It is important after this stage to 'condition' the grain to the required moisture content.

The moisture content of a grain influences its milling ability. As moisture levels increase in the grain, the bran toughens but the endosperm becomes more easily broken down. The ideal occurs at 15·0%–17·5% moisture content and is attained by conditioning at 25°C/77°F for 48 hours or for shorter periods at higher temperatures. At higher moisture levels the bran will stick to the endosperm and not separate easily. Harder wheats are conditioned to higher moisture levels than soft wheats.

The *milling process* itself involves a series of disintegrations followed by sievings or 'siftings'. The first operation in the process is the use of the *break rolls,* usually five sets, which break open the grain. The break rolls

Figure 2.16 Break rolls

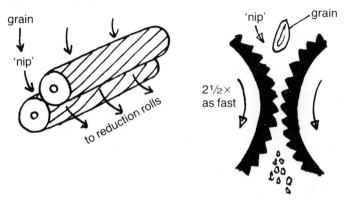

have grooves and one roller rotates 2½ times as fast as the other. As grains pass between the rollers, through the 'nip', one roller holds the grain back effectively as the other roller hits and breaks the grain open, Figure 2.16.

Large particles are produced at this stage but with a little flour. The particles pass to sifters which separate the large particles; these pass to the second break rolls, which operate closer together and have more grooves. Up to five sets of break rolls may be used in this way; see Figure 2.17.

Small particles of endosperm, known as 'middlings' are separated in 'purifiers' and converted into flour by a number of smooth, *reduction rolls*. The reduction rolls are very close together, producing a fine flour, with some starch damaged in the process. This latter fact is important in bread-making, as we shall see later.

The milling operation can be controlled to give flours of varying *extraction rate*. This term is a little confusing as a low extraction rate means a whiter flour and higher rate means the inclusion of more bran and hence a browner flour.

'Wholemeal' flour has had only coarse bran removed and has an extraction rate of 95%. Although the endosperm makes up to 82% of a wheat grain, it is only possible to have a maximum extraction rate of 75% for white flour as bits of bran can contaminate it at higher figures.

White flour always commanded a better price and was in greater demand by the public. However, there is something of a revolution taking place, as a result of the need to eat more roughage, and white flour is declining and wholemeal and other higher extraction flours are becoming more popular. White flours contain little lipid and therefore keep well, whereas wholemeal flours have a shorter shelf-life and have the added problem of containing some *phytic acid* which binds calcium.

Figure 2.17 Wheat milling (simplified)

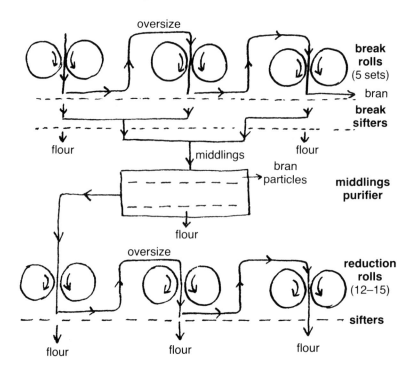

Ageing of wheat flours

Flour used immediately after milling does not produce the best loaves. Ideally, flour should be stored for one or two months to allow improvement to take place, particularly in the wheat proteins. Oxidation of pigments (carotenoids) occurs so that the flour becomes whiter. However, the most significant changes occur in the disulphide bonds (–S–S–) (see Section 1.4.2.1) which increase, joining the protein chains of the flour.

We shall see later how important these groups are in giving the properties to dough of elasticity and extensibility.

Chemicals can be added to flour to improve its baking quality. These *improvers* used at very low levels of only 10–20 ppm are generally oxidising agents. Examples of improvers include potassium bromate, chlorine dioxide, azodicarbonamide, and ascorbic acid (converted in dough to dehydroascorbic acid which is an oxidising agent).

2.4.4 Bread-making

Bread is made from flour, water, salt and yeast. Usually a small amount of fat is also added. From these ingredients a dough has to be made which can be fermented by yeast. The desired dough condition is obtained by mechanical activity, particularly by stretching and folding.

Straight dough system

The ingredients for a 'one-sack' dough would include flour (280 lb/ 127 kg), yeast ($3\frac{1}{2}$ lb/1·6 kg), salt (5 lb/2·27 kg) and water (15–15½ gallons/68–70 litres). The ingredients are mixed with the water at 27°C/ 80°F and the dough is put into a warm place to ferment. After two hours the dough is 'knocked back' by kneading it to even out the temperature and ensure thorough mixing. Any gas (CO_2) is pushed out at this stage. The dough is allowed to rise for a further one hour when it is divided into portions which are roughly shaped. The dough is rested for about 15 minutes, which is known as the 'first proof', and is then moulded into the final shape and placed into tins. The dough is then rested for a further 45 minutes, the 'final proof', during which time it rises in the tin. The tins are placed in an oven, where the dough still rises slightly until the yeast is killed, usually at a temperature of 232°–260°C/450°–500°F for 40–50 minutes. Steam is often injected to produce an attractive glaze on the bread nearer to the end of the baking process, and steam also reduces weight losses by evaporation.

2.4.4.1 Chemical and physical changes occurring in bread-making

When the ingredients are *mixed* they become hydrated with the water and a great deal of the water is absorbed into the mix. The flour proteins become hydrated forming the *gluten* (see the next part for further details). After mixing, fermentation begins and involves the conversion of sugars into carbon dioxide and alcohol by the yeast. Naturally occurring sugars, eg sucrose, glucose and maltose are fermented initially, but the amylases start to break down the starch of the flour to form maltose. The amylases, however, attack the damaged starch, as they find it more difficult to attack starch tightly packed in intact granules. It is, therefore, important to have the right amount of damaged starch resulting from the milling process. Too much damaged starch can allow the production of too many dextrins by α-amylase which lead to a sticky dough, producing denser loaves of poor volume.

The carbon dioxide causes the dough to rise and the gluten retains the gas but also has elasticity and allows the dough's expansion. The production of carbon dioxide and the growth of some lactic acid bacteria results in the dough becoming more acidic.

The 'proving' periods are rest periods in the fermentation process and allow the dough to recover from the effects of cutting and moulding. Unless properly controlled, proving can cause poor texture because of unequal gas production in different parts of the dough. The ideal conditions for proving are 32°C/90°F and 85% relative humidity.

During the first stages of baking, enzyme and yeast activity increase and gas volumes increase because of the temperature rise. These three factors cause the dough to rise rapidly and sometimes to overflow the sides of the tin and occasionally to collapse when there is uneven mixing, resulting in large localized gas bubbles. The yeast activity declines at about 45°C and is killed at 55°C/131°F. However, the amylases are a little more heat-resistant and continue to break down starch up to 70°C/158°F. Physical changes begin to occur rapidly in the dough as the temperature increases, particularly as at 65°C/149°F the starch begins to gelatinize and the proteins coagulate at 75°C/167°F.

Obviously the outside of the loaf reaches a much higher temperature in the oven than the inside, which rarely exceeds 100°C/212°F. This results in the brown crust which is due to the Maillard reaction and the textural changes due to the production of dextrins from starch. The attractive flavour of the crust is as a result of these reactions and from the fact that alcohol is driven from the dough and some is trapped and converted into esters in the crust. The alcohol causes a beer-like smell to emanate from the oven.

The role of wheat proteins in baking

The proteins in wheat are gliadin (40–50%), glutenin (40–50%) and smaller amounts of albumin, globulin and proteose. The wheat *gluten* is a blend of the two protein fractions, gliadin and glutenin. Gluten can be prepared by taking a bread flour and adding 60% of water and allowing it to stand for 30 minutes. Under running water the starch can then be washed away to leave the elastic, somewhat sticky, but tough, gluten.

Gluten can be considered as a system of proteins containing a number of bonding methods. The most significant bonding in operation is the disulphide bridge (–S–S–) between the cysteine amino acid units. As the dough expands there is an interchange between the disulphide bridges (–S–S–) and sulphydryl groups (–SH), and this process releases the pressure as the dough expands. However, as the disulphide bridges reform quickly the dough retains its elasticity, whilst at the same time having this extensibility. Extensibility of the dough may also be due to lipoproteins which form 'slip-planes' in the dough. On baking in the oven the proteins coagulate and with the gelatinized starch form the honeycombed structure typical of the leavened loaf.

2.4.4.2 Rapid dough processes

A number of processes have been developed to speed up the dough-making process but still to produce bread of the normal texture and quality. Basically, the main fermentation processes, which take several hours, are replaced by rapid methods involving oxidation of sulphydryl groups (–SH) and considerable mechanical development of the dough.

The Chorleywood process is used extensively as it produces consistently good loaves, of increased yield, and large amounts of weaker flours can be used. The dough is thoroughly mixed and stretched over a period of 5 minutes, during which time rapid oxidation of the sulphydryl groups occurs. Extra fat, water and yeast are needed. The oxidation of the sulphydryl groups is achieved by using 75 ppm of ascorbic acid. This may be confusing as ascorbic acid is a reducing agent, ie it has the ability to add hydrogen or remove oxygen. However, in the dough the ascorbic acid is converted by enzyme action to dehydroascorbic acid. Dehydroascorbic acid is an oxidising agent and removes the hydrogen from the sulphydryl groups –SH, thus forming the disulphide bridges –S–S– very rapidly.

In this way the dough structure is quickly formed and tightens. The violent mixing carried out in the process corresponds to the gradual stretching achieved during the normal fermentation process. Sometimes the amino acid L-cysteine is added, which tends to make the gluten uncoil a little and expose the sulphydryl groups which can then be oxidised.

Figure 2.18 Action of ascorbic acid in the Chorleywood process

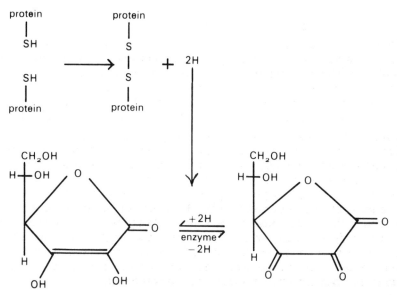

2.4.5 Cakes, biscuits and pastry

Unlike in bread, the gluten network is undesirable in most other baked products. The property of extensibility is often required, however, but without the elasticity. If the shape of a biscuit is cut out then it would be unacceptable for this to contract due to the properties of gluten. In general, flours produced from soft wheats (ie low in protein) yield better cakes than those from hard wheats. Special cake flours called 'high-ratio flours' are available to industry. These specially milled and sieved flours contain more starch and less protein, thus producing the best cakes. The particles in the flour must be carefully controlled and must be as small as is possible.

In addition to flour, *simple cakes* also require as ingredients sugar, fat, eggs and often baking powder. It is not proposed to discuss the various cakes and how to make them, but to review some general principles.

A satisfactory cake can only be produced from a batter in which there has been adequate gas produced and retained. Air can be beaten into the mix and retained with the help of egg and by creaming with suitable fat. Baking powder is often used to produce carbon dioxide in the moist batter and during the earlier stages of baking. Baking powders are formed from sodium hydrogen carbonate (bicarbonate), a slow acting acid and starch. The starch helps keep the other two ingredients dry and the mixture free-flowing. Baking powders often use calcium phosphate $(Ca(H_2PO_4)_2)$ or disodium pyrophosphate $(Na_2H_2P_2O_7)$ as the acid ingredient. Traditional baking powder contained cream of tartar (potassium hydrogen tartrate) and sometimes tartaric acid. These acid constituents react at room temperature to produce carbon dioxide, but some of the gas may escape before baking. Sodium aluminum sulphate has been used as the acid as it reacts only at oven temperatures, but can produce an unpleasant off-flavour. The acidic derivative of glucose, glucono-delta-lactone, is often used in a similar manner.

Air must be whipped into the mix for most cakes, as it influences the final cake volume. The batter must stretch round and retain the air as it is mixed into the mix. The proteins from the flour and eggs do this, but too much protein produces a rubbery cake. This rubbery texture is minimised by the incorporation of fat and sugar into the mix.

Fats interfere with the development of the gluten network and allow the protein particles to slide on each other which leads to a more tender product. Emulsifying agents, such as mono- and diglycerides and lecithin, are added to the fats; sometimes large amounts are added to form 'super-glycinerated fats'. These emulsifiers enable more water to be mixed into the batter, which in turn allows more sugar to be added. As a result of these additions a sweeter, more moist cake is produced, which, because of the higher water content, stales more slowly.

Sugar not only provides sweetness to the cake but is responsible, along with fats, for tenderness. Sugar has the tendency to slow down the development of the gluten. Cakes containing a lot of sugar require more mixing than those with less sugar, otherwise a lower cake volume is obtained after baking.

The amount of moisture in a cake is an important determinant of quality. During the mixing process water dissolves the sugar and hydrates the starch and protein. The protein and starch hold water through hydrogen bonds, even though the protein becomes denatured during baking. When a cake becomes *stale,* moisture is lost and this is due to changes in the starch. A process similar to *retrogradation* occurs (see Section 1.2.2.1) in which hydrogen-bonding of water to the starch alters and water is eliminated from the starch gel.

The high moisture content of cakes can readily lead to mould growth, particularly in wrapped goods. Sodium and calcium propanoates (propionates) will prevent mould growth, and similarly sorbic acid has been found to be useful.

The mixes for *biscuits* and *pastries* are dough-like and are rolled and kneaded, unlike the liquid-like batters for cake-making. In biscuits a large amount of fat is used to ensure disruption of the gluten, so that a crumbly or brittle texture is produced. Baking powder is used but only a small degree of gas production is usually required.

In *pastry* tenderness and a degree of flakiness are fundamental requirements. Ingredients must be mixed carefully and the creaming of the fat must be such that it shows some 'plasticity'. Although the gluten development is again retarded, a layering of the gluten is encouraged to give flakiness to the product.

2.4.6 Some other cereals

Barley

This very hardy cereal is grown extensively for the brewing industry and to a lesser extent for other purposes, such as animal feed.

Malt and a range of malt products are made from germinating barley. The barley is allowed to sprout under moist, warm conditions and thus the α- and β-amylases become active and start to break down the starch endosperm. When the desired amount of maltose has been produced, the grain or the malt is extracted for use in beer making, malt whisky and in vinegar. 'Pearl barley' is grain from which most of the bran and the germ has been removed.

Rye

This cereal is preferred to wheat in colder and most arrid regions as it can withstand adverse conditions. The rye protein contains some gluten but it lacks elasticity and so produces loaves of poor volume and dense texture. Doughs made from rye are unusual in that often they are soured with starter cultures which give acidity and characteristic flavour to the rye bread.

Maize

Maize, or corn in many parts of the world, is popular in many poorer countries, but unfortunately it is also poorer nutritionally. The grain is deficient in lysine and tryptophan and low in free nicotinic acid as most of it is bound and unavailable. The disease pellagra has been common when maize was the staple diet. Boiling maize with alkaline water, such as lime water, releases the nicotinic acid from the bound form, and this is practised in South America and Mexico. Cornflour is an important product from maize and is used extensively in many forms in cooking and food processing.

Rice

It is estimated that over half the world's population depends on rice as food. Milling then 'polishing' rice removes the bran to leave the white grain, and in so doing removes most of the B complex vitamins. Parboiling rice before milling causes migration of thiamin particularly, into the grain so that it is not lost in milling. Parboiling also toughens the grain and makes it less susceptible to insect attack.

Nutrients are lost from the grain during cooking, and so rice should be boiled in a minimum of water so that loss of nutrients is prevented. 'Quick-cook' rice has been pre-cooked to gelatinize the starch, then dried under conditions which cause slight expansion of the internal structure. This gives the product a somewhat sponge-like interior capable of rapidly absorbing water during cooking.

Oats

This cereal is unusual in containing more fat (7%) than other cereals (1–2%). It is also fairly rich in protein, but does not contain bread-making gluten. The fat content can cause problems in becoming rancid during storage. In 1987 oats were shown to lower blood cholesterol levels.

Phytic acid occurs in oats and can combine with calcium, in extreme cases causing calcium deficiency leading to rickets. Soaking the grain for long periods reduces the problem. Oats are used mainly in porridges, but can be used in biscuits and animal food.

2.4.7 Breakfast cereals

Cereal grains and part grains are used extensively in many modified forms as breakfast cereals which are especially popular in Britain. To reduce the consumers' preparation time the starch of the endosperm must be pregelatinized by some method which involves the reduction of the B complex vitamins. Vitamins B_1, B_2, B_6, D_3 and niacin, in addition to iron, are often added so that, it is claimed, many cereals provide over a third of the daily requirements of these vitamins. Three types of breakfast cereals will be discussed below, but the range of products is ever increasing in this highly competitive food sector.

Cornflakes are made from white maize which must be thoroughly cleaned and shelled and must have all bran and germ removed. The white endosperm is broken into pieces which are then steamed for about three hours to gelatinize the starch. A number of additions are made which include salt, sugar, malt and vitamins. The mixture is cooled and then is passed through smooth rollers under tremendous pressure. The flakes produced by this rolling are heated in an oven to toast them and give a somewhat 'bubbly' appearance.

Shredded wheat is made from whole wheat which is added to an equal amount of water and pressure-cooked to gelatinize the starch. The grains are now soft and are shredded by two rollers which produce a continuous flow of strands which are built into layers of the required thickness for the 'biscuit'. The edges of the biscuit are crimped and the product is dried in an oven at about 250°C/482°F for about 20 minutes.

Puffed cereals have been popular for many years and can be made from barley, maize, rye, wheat, rice and even soya beans. In the case of puffed wheat the grains are placed in a special puffing gun into which steam is injected to increase the pressure and cook the grains. The pressure is suddenly released and this causes rapid expansion of the water vapour within the grains causing them to double their size. This process leads to open honeycombed texture typical of this type of product.

2.4.8 Pasta products

Pasta products are made from the Durum group of wheats which are characterized by having tough endosperms which produce a yellow coloured semolina on milling due to carotenoid pigments. Semolina is granular starch from the endosperm of hard wheat and it is better than flour because less water is needed to make the pasta dough, thus facilitating drying of the pasta product subsequently.

There are many different types of pasta and some are enriched with egg or soya flour. Generally there are four stages in the manufacture of these products. The wheat semolina is mixed with salt and water to form a crumbly mixture which is kneaded and worked between heavy rollers to make a dough. The dough must be stiff yet show a degree of 'plasticity' which is obtained after a short resting period. Shaping the dough is performed using a special press with the appropriate die to obtain the right shape and size of the emerging pasta stream. The extruded pasta is cut into standard lengths corresponding to each product type and is then dried in a special drying room, taking up to three or four days. In hot climates, sun-drying has been used for drying pasta, but can take up to two weeks.

Review

1. Cereals

- belong to grass family – *Gramineae*
- most have similar chemical compositions
- composition varies with variety, weather, growing conditions
- moisture level 10–14%, protein content 7–13%
- all cereal grains have: an embryo, endosperm and bran
- vitamins, particularly B complex, found in bran

2. Wheat milling

- 'Strong' wheats have higher protein content – good for bread
- 'Weak' wheats have lower protein content – good for cakes and biscuits
- milling divided into:
  (1) cleaning and conditioning
  (2) size reduction and separation
- moisture content important in milling – condition to 15–17½% moisture
- low extraction rates – up to 75% – mean white flour
- high extraction rates – up to 95% for wholemeal flours
- flour must be aged to improve bread-making properties.
  Improvers, usually oxidizing agents, added to accelerate ageing

3. Bread-making

- bread made from flour, water, salt, yeast and some fat
- straight dough method – relies on wheat to produce CO_2 which stretches dough to produce honeycombed bread texture
- Chorleywood process uses rapid mechanical development of dough plus rapid oxidation of sulphydryl groups (–SH) and form disulphide bridges (–S–S–).
- chemical and physical changes –
  (1) during mixing protein and starch become hydrated
  (2) gluten formed from gliadin and glutenin

(3) fermentation converts sugars to alcohol, then α- and β-amylases attack starch to release maltose which is fermented

(4) gluten retains CO_2 produced, but shows elasticity and extensibility –SH groups converted to –S–S– bridges

(5) during baking starch gelatinized and protein coagulated

(6) crust formed – colour due to Maillard reaction

4. Cakes, biscuits and pastry

– do not require gluten network
– use 'weak' flours of lower protein content
– 'high ratio flours' – small starch particles and low in protein
– baking powders used as raising agents:
  sodium hydrogen carbonate, a slow acting acid and starch
– air must be whipped into batter
– 'Superglycerinated' fats contain more emulsifying agents – allow more water in a cake batter and therefore more sugar – results in a sweeter, longer lasting cake
– in biscuits more fat used to disrupt gluten and produce crumbly texture
– in pastry gluten restricted and layered to produce flakes

5. Other cereals

– Barley – used to produce malt for beer
    – must be allowed to germinate to activate amylases to break down starch
– Rye – some gluten – produces poor loaves
    – dough often soured by starter cultures
– Maize – (corn) little free nicotinic acid
    – released by cooking with alkaline water
    – cornflour main product
– Rice – par-boil to cause vitamins to migrate to endosperm from bran
– Oats – higher in fat than others and contain more phytic acid
– Breakfast cereals – whole or part grains
    – starch pre-gelatinized then puffed, shredded or flaked
– Pasta products – made from Durum wheat –
    – tough endosperms producing semolina

Practical exercises: *Cereals*

1. Microscopic examination of dough

Add a small amount of water to a bread flour. Knead and work into a dough. Take a small amount and make a thin slide preparation. Cover with a cover-slip and place a drop of iodine next to the cover-slip. Starch will be stained and gluten will clearly be distinguished.

167

2. Separation of gluten

Make a dough as in 1. Cover with tap water for half-hour. Place dough in muslin and work between fingers under a running tap to eliminate starch. Gluten remains in the muslin.

3. Water absorption

Weigh 20 g of flour into a dish and add water from a burette. Mix the flour and water and note the amount of water used to produce a dough of normal bread consistency. Compare the water absorption of different flours.

4. Yeast

Examine a suspension of fresh yeast and dried yeast under the microscope. Try to observe the multiplication of the yeast cells by 'budding'.

5. Comparison of bread-making potential of flours

Make a mixture of 40 g flour, 60 cm^3 of water and 3 g of yeast. Pour into a measuring cylinder (250 or 500 cm^3). Note initial level. Observe the rise in the 'dough' as the yeast fermentation proceeds. Note the height achieved by each and conclude which flour is best for bread-making, taking into account its elasticity and retention of gas.

6. Flour additives

Test samples of flour for:

(a) ascorbic acid:

add 1 drop of a solution of dichlorophenol indophenol (DCPIP) to a suspension of flour in dilute acetic acid. The pink colour will be decolourised by ascorbic acid.

(b) potassium bromate:

Make a paste of flour in water and add some potassium iodide in hydrochloric acid. The bromate will be coloured black.

2.5 Beverages

The body needs to take in a large amount of water daily, between 1 and 2 litres. This water is taken in as food, many food products being at least 80% water, and in the form of drinks. Beverages are not normally consumed for their food value but for thirst-quenching and often for their stimulating effects!

It is convenient to divide beverages into those which are non-alcoholic, eg tea, coffee, cocoa and carbonated soft drinks, and alcoholic drinks which include wine, beer, cider and spirits.

2.5.1 Non-alcoholic beverages

2.5.1.1 Carbonated drinks – mineral water

Artificially carbonated water was first produced in the 18th century and has been popular ever since. Carbonated drinks are sweetened, flavoured, acidified, coloured, artificially carbonated and often chemically preserved. The carbon dioxide was originally obtained from sodium carbonate or hydrogen carbonate by the action of a weak acid in the drink. As sodium salts were used, the name 'soda' was adopted and can still be seen today, eg strawberry cream soda.

Sweetening of these drinks has been traditionally carried out using sucrose as a colourless syrup to give the finished drink about 12% sugar on average. Sugar also has the advantage of giving 'body' and some degree of mouth-feel to the drink. There has been an enormous increase in drinks with reduced sugar levels for slimming diets. Sucrose is significantly reduced or totally replaced by saccharin, or any other approved sweetener such as aspartame*, and the body of the drink is produced by adding a small amount of pectin or carboxymethyl cellulose.

Carbonated drinks must have an attractive flavour which must also be stable under acid conditions in the drink and unaffected by light entering the bottle. Natural flavour extracts do present some problems, as for example, terpenes change in the presence of acid. A small amount of emulsifier is often added to prevent separation of essential oils on standing. Synthetic flavours offer many advantages to the manufacturer and are often used.

Synthetic *colours,* particularly coal tar dyes, are used for similar reasons and in cola-type drinks caramel is used. Natural colours generally do not possess sufficient pigment depth and are often not stable in the acidic drinks.

The *acidity* will in part come from the dissolution of carbon dioxide in the water to form carbonic acid.

$$CO_2 + H_2O \rightleftharpoons H_2CO_3$$

* Breaks down under acid conditions giving a slightly shorter shelf-life.

Further acids are used to enhance particular flavours, citric, tartaric and malic acids are common, but phosphoric acid is used in cola-type drinks. The final pH is about 3. Although acids help preserve the drink, long-term keeping-quality is only obtained by using preservatives, usually benzoic acid. Often sodium benzoate is used as this breaks down to release benzoic acid in the acidic drink.

Carbonation of the drink is achieved by exposing a large surface area of the cooled drink to pressurised jets of carbon dioxide. The gas is more soluble in water at lower temperatures than at higher temperatures. After carbonation the drink can be filled into bottles or cans.

2.5.1.2 Tea

Tea comes from only the young leaves of the evergreen tea bush, which yields suitable leaves after three years and up to 50 years afterwards. Like coffee, tea is a stimulating drink as it contains caffeine. (Figure 2.19.)

Figure 2.19 Caffeine

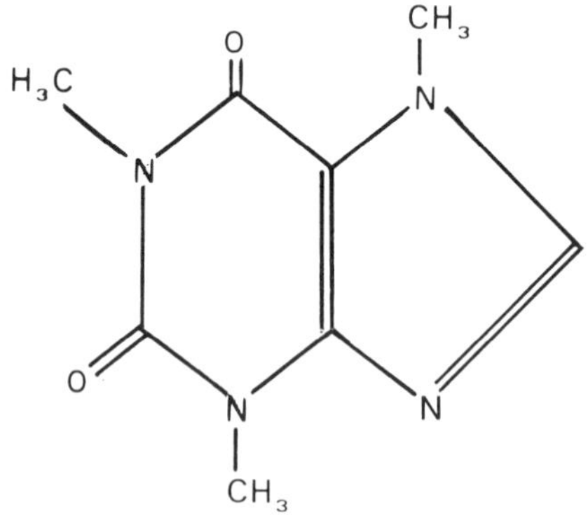

The colour and strength of the tea, together with some body and often considerable astringency are due to the *tannins* in the tea. Tea aroma is produced by a small amount of essential oil in the best and most expensive types of tea. Tea taken with a meal has been shown to reduce the absorption of iron by as much as 64%.

The *processing* of the tea can produce three main types – green, black or oolong. Enzyme activity (polyphenolases) is responsible for black tea and partly for the lighter oolong tea, but the enzymes are inactivated in green tea.

To produce *black tea* the leaves must wither and soften and are partially dried. The cell walls are ruptured by passing the leaves between rollers to release the juices and thus activate the enzymes. The enzymes are active for up to five hours before they are inactivated by drying at about 93°C/200°F. Finally the moisture content is brought down to below 5%.

To produce green tea the enzymes must be inactivated usually by steaming the leaves. The leaves are then simply rolled and dried.

Oolong tea is intermediate in being partially steamed, thus allowing some polyphenolase activity.

2.5.1.3 Coffee

Coffee trees are also evergreen, but produce fruit after 5 years, which when ripe are red and often called 'cherries', see Figure 2.20.

The 'cherry' contains two coffee beans enclosed in a tough skin. To remove the beans, the ripe cherries are passed to pulping machines which remove the pulp of the cherry to leave the two beans surrounded by a coating of mucilage. This mucilage is removed by natural fermentation of the beans or special washing techniques. The beans are dried to a moisture content of about 12% and undergo flavour and colour changes during the process.

The dry hulls of the beans are removed by friction and the beans are carefully sorted and graded for size, colour and potential brewing quality.

The flavour of coffee is further developed during *roasting* which is carried out at about 205°C/400°F for 5 minutes. Essential oils are given off and give the fragrant aroma of roasted coffee. The coffee is ground down to size depending on its intended use as different particle sizes affect brewing time.

Figure 2.20 Coffee 'cherry'

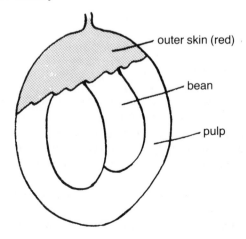

In brewing coffee, about one third of the constituents are extracted by hot water. Caffeine is extracted easily and almost completely after 10 minutes brewing; however shorter brewing times are better to avoid extracting bitter phenolic compounds. The composition of coffee is complex, and over 200 compounds have been identified. As well as caffeine, chlorogenic acids and nicotinic acid derivatives are important.

Instant coffee is made by dehydrating the brew from large percolators. Generally spray-drying is used, but high quality coffee is made by accelerated freeze-drying (see Section 4.3.1). However, even the best instant coffee can lack the full flavour of freshly brewed coffee. Volatile constituents can be trapped and recovered during roasting, grinding and extraction to be added back to the coffee.

Coffee consumption has attracted attention as a possible health hazard. Studies have shown possible connections between coffee and coronary heart disease, high blood pressure, some forms of cancer and diabetes. In most cases results are inconclusive. However, in Norway in 1983, total blood cholesterol and triglyceride concentrations were found to increase with increasing coffee intake. This so-called Tromso heart study has shown that this relationship should be investigated further. During 1987 sales of decaffeinated coffee grew enormously; the caffeine being removed by steam or solvent treatment of the beans.

2.5.1.4 Cocoa and drinking chocolate

Cocoa, like tea and coffee, contains some caffeine, but also it contains another stimulant, theobromine. Cocoa beans grow in pods, with 25 or so beans arranged in rows in each pod. The beans are surrounded by mucus and pulp which is removed by fermentation carried out traditionally by heaping the beans and covering them with leaves. During fermentation colour and flavour are developed. The beans are dried to about 7% moisture and later are roasted to develop their flavour further. Special crushing machines and 'winnowers' are used to separate the grain and hull from the remainder known as the *nibs*. The nibs are passed through various mills where they are torn apart and ground releasing fat from the cells. The heat of the grinding process melts the fat and this liquid product is known as *chocolate liquor,* the essential ingredient of chocolate. After removing the chocolate liquor the cake left is prepared for cocoa powder, which is lower in fat. The cocoa powder may be treated with weak alkali, eg potassium carbonate which improves the solubility of the cocoa and darkens the colour. Often sugar, vanilla essence and salt are added to cocoa to produce a typical drinking-chocolate.

2.5.2 Alcoholic beverages

2.5.2.1 Wine

Grapes have been fermented by yeasts to make wine since at least 4000 B.C. Although other fruits can be used to make wine grape is the most suitable. Yeasts naturally occur on the surface of grapes as bloom and will readily ferment the crushed grapes. Wine production, therefore, can be easily carried out with the minimum of equipment. However, the enormous demands for consistent and high quality wines has led to elaborate procedures being developed.

The quality and characteristics of a wine depend on: variety of grape, climate, soil, method of manufacture and maturation.

General characteristics of wines

Colour

Red wines are produced from black grapes, eg *Pinot noir,* which are pulped and fermented with the skins present. The pigments of the skins, mainly anthocyanins, are extracted and as alcohol is produced during fermentation this extraction is accelerated. The longer the skins remain in the fermentation the darker the colour. The pulp is strained off and then fermentation proceeds to completion.

Rosé wines are produced in a similar way, but the skins only remain for about 24 hours before straining.

White wines can be made from most grapes, even black ones, as the flesh of the grapes is seldom coloured. The grapes are crushed and the resulting pulp is hydraulically pressed (formerly this was done with the feet) to separate juice from the pulp. A yeast starter culture is added to ferment the juice, which is often treated prior to this with sulphite to kill 'wild' yeasts and bacteria.

Alcohol content and sweetness

Obviously, the sugars of the juice are fermented to produce alcohol and in so doing the sweetness of the juice decreases. If all the sugars are fermented, a *dry wine* results, and the alcohol content will be between 9–14% usually. In warmer climates the grapes will fully ripen and contain more sugar, which leads to a wine with more alcohol and possibly one which is sweet. Yeasts generally can only ferment up to about 14 or 15% alcohol before they are inhibited then killed by the alcohol.

Some wines, such as Sauternes, are very sweet because the grapes used have been 'attacked' by a mould – *La pourriture noble* – the noble rot. This mould concentrates the sugars in the grapes, making them almost raisin-like, and also the mould produces glycerol which adds to sweetness.

Fortified wines have alcohol contents of 17–21%. To achieve this, spirit, usually brandy of lower quality, is added to fortify a normal wine.

Sparkling wines

During fermentation carbon dioxide is produced; in a number of sparking wines this gas is retained to give the bubbles. Special alcohol-tolerant strains of yeast are used and residual sugar is involved in a secondary fermentation in the bottle. Champagne is the best sparkling wine as it has small, long-lasting bubbles.

Semi-sparkling wines, eg Mateus Rosé, are produced by a special fermentation, the malo-lactic fermentation. Malic acid is broken down, by certain bacteria, to give lactic acid and liberate carbon dioxide, as shown in the following equation:

$$CH_2COOH$$
$$|$$
$$CH(OH)COOH \longrightarrow CH_3CH(OH)COOH + CO_2$$

malic acid   lactic acid

Wine-making

The yeast used for wine-making is a variety of the common yeast *Saccharomyces cerevisiae* and has an ellipsoidal shape, hence *S. cerevisiae var. ellipsoideus.* The yeast acquires individual characteristics when growing on different grapes in different areas, for example, the ability to produce more alcohol. The grapes are crushed and an active yeast culture ferments the sugars until either: all the sugar is used; the yeast is poisoned by high alcohol levels; or additional alcohol is added. Sometimes benzoates can be added to stop further fermentation.

After fermentation, 'racking' is carried out to separate the wine from the sediment, particularly dead yeast which starts to break down. The wine may be aged in casks or tanks for several months, or even years, for heavy red wines. Red wines require a long maturation period to break down their high content of tannin. After maturation, the wine is filtered and stabilized by the addition of benzoate or sulphites. Crystals of salts of tartaric acid are often removed at this stage. Some wines are pasteurized before bottling.

During 1986 and 1987 some Austrian and German wines were found to be sweetened with an industrial solvent, diethylene glycol* to give the appearance of a better wine.

2.5.2.2 Beers, ales and lagers

Beer has been made for thousands of years, but consumption now exceeds that of soft drinks. Beer-making is more complex than wine-making, as fermentable sugars must be extracted from grain, particularly from barley. Malting of barley must be carried out to

* Anti-freeze is ethylene glycol.

produce maltose and other fermentable sugars (see Section 2.4.6). Other grains can be used to extend the malt, since the amylase content of malt is capable of breaking down more starch than is present in barley itself. These grains, usually boiled, modify the flavour of beer and minimise the formation of protein hazes.

The malt and cereal grains are roughly milled and hot water is added in a large container – the mash tun. The temperature is controlled at about 65°C/149°F, so that amylase activity is rapid. This *mashing* process takes about three hours, after which the liquid, *the wort*, which is rich in dissolved sugars, is allowed to drain from the grains. The wort is passed to a copper where it is boiled and hops are added. The hops add flavour, bitterness, natural preservatives, and protein coagulants to the wort. The wort is then filtered and rapidly cooled. Fermentation is carried out at about 15°C/59°F.

In typical British beers the carbon dioxide produced during fermentation carries the yeast to the top, and is called top-fermenting. In lagers, the yeast is bottom-fermenting. Finings may be added to clarify the beer, or it is filtered to remove yeast, then bottled or kegged. Carbonation is carried out by sugaring barrels in a traditional process or by injection of carbon dioxide.

Compared with wines the alcohol content of beers is lower at about 2–5%. Beers which are heavier with more 'body' are richer in carbohydrates and proteins. These beers are also richer in minerals and are good nutritionally.

2.5.2.3 Spirits

Distillation is the only method of producing drinks with alcohol concentrations above 20%. Distillation is dependant on the difference in volatility between water and ethanol. In heating a wine or beer both water and alcohol are distilled, but as alcohol boils at 78°C, compared with water at 100°C, the alcohol concentrates in the vapour and must be trapped and condensed. This is a difficult process for a number of reasons. Rapid heating during distillation can cause loss of alcohol and unwanted chemical reactions. During the initial stages of distillation, methanol, which is toxic, can be given off and therefore the 'spirit' collected at first must be discarded.

The raw spirit must be aged over long periods in wooden barrels. Some oxidation occurs producing flavours and some flavour is picked up from the cask, ideally made of oak.

Liqueurs are produced by adding a special flavouring, such as an essential oil, to a spirit and usually syrup for sweetening.

Review

1. Carbonated drinks

- carbon dioxide from sodium carbonate and weak acid – hence 'sodas'
- sucrose used for sweetening and to give body
- in dietetic drinks – saccharin or aspartame used and carboxy-methyl cellulose to give body
- natural flavours and colours inferior to synthetic versions
- acidity from dissolved CO_2 and other acids, such as citric, and phosphoric in 'cola'

2. Tea

- young leaves of tea bush
- contains caffeine
- can inhibit over 60% of iron absorption in a meal
- colour, strength, body and stringency due to tannin
- three types, green, black and oolong
- black and oolong need active enzymes (polyphenolases) to give colour

3. Coffee

- coffee beans enclosed in pulp and skin as a 'cherry'
- fermentation required to remove mucilage
- drying develops colour and flavour
- roasting further develops flavour
- constituents, up to ⅓, extracted by brewing
- instant coffee made by dehydrating brew by spray-drying or AFD
- coffee attracts attention as possible health hazard – some correlation between coffee consumption and increases in cholesterol in blood

4. Cocoa

- contains less caffeine, but also stimulant, theobromine
- chocolate liquor used for chocolate manufacture, remaining cake used for drinking chocolate

5. Wine

- quality and characteristics depend on variety of grape, climate, soil, method of manufacture and maturation
- red wines produced from black grapes by fermenting pulp and skins
- rosé wines only short period of pulp fermentation
- white wines press out juice and ferment in absence of skins
- alcohol content depends on sugar content, type of yeast and presence of moulds which can concentrate juice within grapes, eg Sauternes
- sparkling wines require secondary fermentation in bottle, eg Champagne
- semi-sparkling wines depend on malo-lactic fermentation
- fortified wines have added spirit, eg brandy

6. Beer

– more complex than wine-making
– must extract fermentable sugars from grain, eg barley
– barley must be malted to produce sugars
– hops add bitterness, flavour, preservatives and protein coagulants
– British beers – top-fermenting yeasts
– lagers – bottom-fermenting yeasts

7. Spirits

– must distil and collect alcohol carefully
– raw spirit needs long maturation
– liqueur: spirit with added syrup and flavour, such as essential oils

Practical exercises: *Beverages*

1. pH and titratable acidity

Compare the pH of a range of juices and soft drinks, using a pH meter (if available) or pH papers. Titrate 10 cm^3 samples of drink with 0·1 M sodium hydroxide, using phenolphthalein as indicator. A light pink colour indicates the end-point, record volume of hydroxide used.

2. Saccharin identification in drinks

To a sample of drink, add a small amount of resorcinol. Add a few drops of concentrated sulphuric acid (CARE!). Heat in a water bath until a green colour is produced. Cool completely, then add some water and sodium hydroxide solution. A fluorescent green is produced, indicating the presence of saccharin.

3. Effects of pH

To a coloured drink, for example blackcurrant juice, add varying amounts of acids and alkalis. Record the pH and note the colour change.

4. Effects of pH on tea and coffee

Prepare brews for tea and coffee, and filter. Repeat 3.

2.6 Chocolate and confectionery

2.6.1 Chocolate

We saw in Section 2.5.1.4 how cocoa was processed eventually to produce *chocolate liquor*. This liquor is about 55% fat,17% carbohydrate and 11% protein. In addition it contains about 6% tannin and 1·5% theobromine, a stimulant similar to caffeine. The presence of this substance explains the slightly addictive nature of some chocolate.

On cooking this liquor a bitter chocolate is produced which can be used in baking. Further processing is necessary with sugar to yield sweet chocolate and with sugar and milk to produce milk chocolate.

Cocoa butter is the fat which is removed from chocolate liquor and in many ways it is an unusual fat. Cocoa butter is quite hard at normal temperatures but melts quickly at just under body temperature. This characteristic enables chocolate to be made into various shapes and bars and to be used as coatings for sweets and biscuits, where it remains solid until it is eaten, when it softens readily in the mouth. Processing of chocolate is designed to ensure that the cocoa butter crystallizes in a stable form in the chocolate.

There are three unstable forms which can lead to poor texture and the formation of white 'bloom' on the surface of the product.

Plain chocolate is made from chocolate liquor, or cocoa nibs, to which is added cocoa butter and some sugar. Cocoa beans must be carefully selected and roasted to give the distinctive flavour of plain chocolate. The ingredients are mixed in a special mixer which has scraper blades and heavy crushing rollers. An even paste is produced in this mixing process and is ground to reduce the particle size until it is completely smooth. Large steel rollers carry out this grinding process and are referred to as *refiners*. A film of chocolate is passed from one roller to the next and there are usually five in all. Often plain chocolate is refined several times before the final process of *conching*.

Conching is performed in a large metal vat with a large roller and is a slow process, sometimes taking up to two days at 65°C/149°F. The warm chocolate from the refining process is poured into the vat and extra cocoa butter and some vanilla flavouring are added. The roller moves backwards and forwards in the vat moving the chocolate continually. In this manner the chocolate is helped to develop its full flavour and aroma, and the right texture.

The chocolate can now be cooled and moulded. However to obtain a glossy finish the chocolate is cooled until it thickens slightly and then is warmed again. This process is *tempering*. The warm, tempered chocolate flows into moulds where it cools and sets in a controlled manner.

Milk chocolate contains milk in addition to the constituents of plain chocolate. Milk must be condensed under vacuum and usually sugar is added to produce a sweetened condensed milk which is mixed with chocolate liquor and dried to produce a powdery solid. This powder is made into chocolate by being mixed with cocoa butter and refined and conched as for plain chocolate.

2.6.2 Confectionery

A vast range of confectionery products is produced based on the ability to manipulate sugar, the principal ingredient. This manipulation is accomplished by controlling the state of crystallization of the sugar and the sugar:moisture ratio. The sugar may be in the form of large or small crystals or it may be non-crystalline and glass-like. The mixture may be hard or soft according to the moisture level and air may be whipped into the product. If the sugar is *crystalline* it may be in the form of one crystal forming the whole product, such as in 'rock'. Very small crystals exist in fudges and fondants, eg chocolate fillings.

Non-crystalline sugar forms hard or brittle, or chewy sweets and gums. The sugar is amorphous and the water content varies from about 8% and up to 15% to produce the softer products. Marshmallows are very soft as air is whipped into them.

If sugar is added to water in a ratio of 2 : 1, a final solution of 66% sugar results which, on cooling, becomes *supersaturated*. Upon further cooling and with some agitation the sugar crystallizes. This crystallization can be accelerated by 'seeding' with just one sucrose crystal. Very high concentrations of sugar may solidify as an amorphous mass, for example in the manufacture of boiled sweets.

Invert sugar is an important ingredient in many sweets as it is sweeter than sucrose but also controls its crystallization. Invert sugar encourages the formation of small crystals essential to the smoothness of many products such as fondants, fudges and soft mints. Corn syrup and special fructose syrup are now often used.

Caramelization is an important process in the production of a number of products. If sucrose is heated, either as a solid or solution, it undergoes decomposition to produce a range of brown products collectively known as *caramel*. Too much heating will produce a bitter and very dark 'caramel'. Often in the production of caramels milk powder is added in a small amount to facilitate the Maillard reaction, rather than just allowing the process of caramelization to proceed alone.

179

Review

1. Chocolate

– chocolate liquor: 55% fat, 17% carbohydrate, 11% protein
– contains stimulant, theobromine
– cocoa butter – fat from chocolate liquor: unusual, hard at room temperature, but melts just under body temperature
– unstable crystal forms of cocoa butter can cause white bloom on chocolate
– conching helps develop flavour and aroma and with tempering develops right texture
– milk chocolate – added condensed milk

2. Confectionery

– control of sugar crystal size and ratio of sugar:moisture
– crystalline products –
  large crystals – rock
  small crystals – fudge, fondant
– non-crystalline products – boiled sweets, butterscotch, gums
– caramelization must be controlled to prevent bitter and dark coloured products
– invert sugar important to add sweetness and control sucrose crystallization

Practical exercises: *Confectionery*

1. Caramelization of sugar

Heat sugar in an ignition tube, note production of brown colour and typical smell. On further heating note the production of unpleasant and acrid odour of burnt caramel.

2. Inversion of sucrose

Make a solution of sucrose and test with the Fehling's test. Add 1 cm^3 of dilute sulphuric acid to a second solution of sucrose and heat in a water-bath. Test with Fehling's solution and note the positive result due to the production of invert sugar.

3. Crystallization of sucrose

Crystallize sugar in a number of ways to make crystals of different sizes and different products:
  (i) make a strong solution of sugar in water by heating, cool and allow to crystallize
  (ii) repeat (i) but seed with small crystals of sugar
  (iii) repeat (ii) but stir
  (iv) repeat (ii) and add a small amount of cream of tartar.

2.7 Product development

A business depends for its continued existence on its customers. Customers require a varied diet and a lot of choice. New food products, therefore, are continually needed.

Food products pass through a type of 'life-cycle'. Some products become established and sell in the same form for many years, whereas others, such as chocolate bars, quickly come and go. In Figure 2.21 each stage of the life-cycle can be seen.

Figure 2.21 Product life cycle

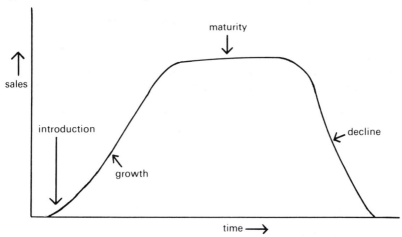

Product sales start slowly in the *introductory phase,* then rapidly grow during the *growth phase,* especially when extensively advertised. Sales eventually level out during the stage of *maturity* and finally *decline.* Falling sales are often rejuvenated by special offers, free gifts and addition of new flavours and colours. Ailing food products carry high factory costs and quite often cannot be rejuvenated successfully. It is often better to launch a new food product, and something like forty new products are launched in the UK each month.

Many food products fail to sell for more than eighteen months. The reasons for these failures, often in the region of 90%, are varied and obscure. Fortunately the work of home economists, food scientists and technologists is rarely the cause of product failure. Failure often lies in the marketing of the product, sales techniques and advertising. Economic factors affect food sales, often quite quickly. Expensive convenience foods and delicatessen products tend to decline in slumps and times of high taxation.

A number of factors are constantly tending to promote the development of new or improved food products, and these factors can be divided into market and factory influences.

A number of factors which are recognised during the *marketing* of an existing food product can indicate the need for new products. *Market research* can discover a new consumer 'need' for a particular convenience food product (see Section 3). *Consumer complaints* could indicate that the present product is inadequate. However, it is estimated that only about 10% of consumers complain about a faulty food product.

Often further products might be needed to *expand* a range of products, for example more flavours may be added in a dessert range. A new product should take into account the company's strengths: design and development ability; production facilities; customer representation; distribution system; and marketing skills.

A *competitor* might produce a better or cheaper product, thus forcing the re-development of other existing products. Some companies develop new foods with novel ideas, others just follow and produce 'me-too' products, but at a lower price.

There are often a number of factors which can arise in a *factory* which necessitate the development of new products, range extensions and improvements in existing products. *Processing plant* might be under-utilized, particularly in seasonal food processing. New products could even out *seasonal variations* and their effects on the factory and employment.

Some food processes produce a lot of potentially valuable *waste products*. These waste products can be used to make new products, even animal foods, such as dog biscuits from cereal waste.

An existing food product, due to increases in raw material costs for example, may become too expensive to produce. Costs could be reduced by developing and producing food products which are easier or cheaper to make.

Pure research within a research establishment or university might provide new product ideas which can be developed into new food products which can be marketed and sold at a profit. Similarly the advent of new technology, such as the production of single-cell proteins, can lead to development of new ranges of foods.

Food products can be developed and produced but only within certain constraints laid down in *legislation* and agreed *codes of practice* within the industry. There are three main Acts controlling food. The *Food and Drugs Act* concerns general food quality; the *Weights and Measures Act*, the declared weights of the product; and the *Trades Description Act*, the labelling and description of the product. There are a number of *regulations* concerning individual aspects of food such as; food additives, food contamination, compositional standards, hygiene and labelling. These Acts and regulations are enforced by law and a food company must comply with them when developing and manufacturing a food product. There are also a number of *Codes of Practice* which are agreements between firms, or the government, to manufacture something in a particular way, or to include a particular constituent at a certain level. Some codes are international such as the Codex Alimentarius.

People have to eat. In Britain, with static population growth, variety is the problem. In developing countries feeding people is the problem. Above all, new food development must be aimed at cutting down waste.

This is a brief look at product development which is a fascinating area of work offering immense opportunities for food scientists, technologists and home economists.

Section 3

Handling and Preparation of Food Raw Materials

This section of the book covers the storage of raw materials before processing and preservation, and handling and preparation of these materials to make 'convenience' foods. All operations required to be carried out before preservation are covered here.

Convenience foods may be defined* as foods for which the degree of culinary preparation has been carried to an advanced stage by the manufacturer and which may be used as labour-saving alternatives to less highly-processed products. This preparation of the food means that the product is easier and quicker to use than the basic product available to consumers. Foods are convenient for different reasons. They may be convenient to buy, being more economical in the long run with less wastage; they may be convenient to carry home, one tin of dried milk will make perhaps three litres of liquid milk. Convenience in storage is an important factor as canned and packeted foods easily store in a cupboard or pantry and most homes now have a freezer to store frozen goods. The speed and convenience of preparation are probably the most important factors in evaluating a convenience food. The old type of field-dried peas (Marrowfat) are a convenience food in that they are sorted, cleaned and pods are removed; however they may take all night to rehydrate, which is hardly convenient! New methods of drying peas can produce a product which rehydrates in a few minutes. Packaging a food into convenient serving sizes is another aspect of convenience. Fully-prepared convenience foods are relatively new and in these foods convenience should be in their completeness.

The food factory takes in a large quantity of food material which must be stored until required. The food material is then prepared in a number of ways which include: cleaning, sorting, grading, size reduction, mixing, concentration, filtration and blanching. Following those of these procedures which are required, the product is preserved and packaged.

*National Food Survey Committee

3.1 Storage of raw materials

In Section 2.3.3 the storage of fresh fruit and vegetables was discussed. Factors controlling the quality of stored produce were shown to include temperature, use of gases, reduction in pressure and the specialized use of coatings. These methods can be applied to the storage of a wide range of food products, and even the use of special coatings might have applications in other products yet to be investigated.

3.1.1 Chilled storage

In 1982 the Institute of Food Science and Technology in its *Guidelines for the handling of chilled foods* defined the term 'chilled food' as: 'a perishable food which, to extend the time during which it remains wholesome, is kept within controlled ranges of temperature above $-1°C$ and below 8°C. Since 1 April 1991 this has become a legal requirement. Chilling cannot preserve a food indefinitely but can reduce spoilage caused by micro-organisms and enzymes. Micro-organisms grow in certain temperature ranges and each type of organism has an optimum temperature for rapid multiplication (ie growth).

Thermophilic organisms, usually bacteria, grow best at 50–70°C/122–158°F and they are usually very heat-resistant. Chilled storage will completely inhibit these organisms. The *mesophilic organisms* have optimum growth in the range 20–40°C/68–104°F and are common spoilage organisms of food. Generally they are inhibited in chilled stores, but occasionally some can grow at 5°C/41°F. The *psychrophilic* organisms are low temperature organisms growing rapidly at 0–10°C/32–50°F and even below 0°C if water is available. This group of micro-organisms includes a number of bacteria which readily spoil chilled foods.

Enzymes are similar to micro-organisms in having optimum temperatures or ranges for their activity. In Figure 3.1 the typical shape of a curve showing enzyme activity against temperature can be seen. Like micro-organisms some enzymes are active at chilled temperatures and cause changes in the product particularly in flavour, colour and texture.

It is essential to chill food as quickly as possible and keep it chilled. Chilled foods must be handled hygienically and with care to avoid damage such as bruising.

Chilling is **not** freezing, as in the latter case water in the food is frozen and the temperature of the frozen food will be well below freezing point.

During *storage* chilled products gradually deteriorate and any quality

Figure 3.1 Enzyme activity and temperature

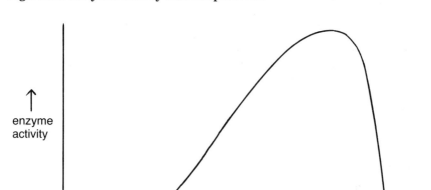

loss cannot later be recovered during processing. The storage of chilled products, whether in a factory or in the home, is subject to certain commonsense rules. Raw meat, fish or poultry should be stored so they do not drip on to other foods, particularly those which have been cooked or processed and will be eaten without further cooking. Leafy vegetables should be covered or packed in such a way as to control the loss of moisture by evaporation. Prepared salads such as cole-slaw can also dehydrate readily. The domestic refrigerator is similar to the factory chill-store, but on a smaller scale, and generally maintains a temperature of 7°C/44°F or over.

Although many raw materials are stored in a chilled form, prior to further processing, there is a large sale of fresh chilled foods which have not been processed further. The *Food Labelling Regulations (1980)* require chilled foods, among others, to be labelled with either, 'sell by' or 'best before' dates on the package. The 'sell by' date is the latest recommended date of sale of the food, and also includes an indication of the period after purchase for which the food will retain its freshness, if stored according to the instructions supplied. The 'best before' date is the date up to which the food can be expected to retain its freshness and eating qualities. It must be remembered that these dates are approximations, as the food does not suddenly become unpalatable but gradually deteriorates.

3.1.2 Controlled atmosphere storage

Gas storage, in addition to chilling, can greatly extend the storage life of many food materials. Controlled atmosphere (CA) and modified atmosphere (MA) have been discussed in Section 2.3.3.2 for use with fruits and vegetables. Other products can benefit from the addition of about 10% carbon dioxide to the store, as it is particularly effective against moulds and some bacteria. Chilled beef will keep for longer periods if 10–15% carbon dioxide is added to the store. Higher concentrations, however, cause the meat to lose its bright red colour due to the loss of oxymyoglobin in the diminished concentration of oxygen. Myoglobin and possibly brown metmyoglobin is produced which changes the meat to dull red and brown.

Eggs can be stored with only about 2½% added carbon dioxide. This reduces the loss of carbon dioxide from the eggs and the corresponding rise in pH is inhibited, thus retarding changes leading to lower quality.

Hypobaric storage (low pressure) has been used for a range of products. However, as discussed in Section 2.3.3.3 it is very expensive and similar results can be obtained using cheaper CA methods.

3.1.3 Frozen storage

Food products must be frozen quickly and maintained at an even, low temperature during storage. The freezing techniques and the effects on food quality are reviewed later in Section 4.2.

In the factory the cold-store corresponds to the domestic freezer. After being frozen quickly by some freezing technique (Section 4.2) the product must be stored in a cold-store at the recommended temperature. Ideally, the lower the temperature of the store the better the quality of the product after thawing. A temperature of −29°C/−22°F has been found to give better results with most food materials, compared with a temperature of −18°C/0°F, which used to be in widespread usage.

If the temperature of storage is allowed to fluctuate a number of problems can develop. At −18°C/0°F it has been found that a temperature fluctuation of ±3°C caused green vegetables to start to change to a dull greyish colour due to pheophytin production from chlorophyll (see Section 1.7.1.2).

Similarly mince beef was found to start to become rancid. Fluctuating temperatures also cause loss of weight of the product as ice will sublime to water vapour. If the product is well packed the water vapour may condense and re-freeze on the inside of the packaging material, producing frost which falls out when the package is opened. This is frequently seen in domestic freezers and with products such as frozen bread. In exposed areas of a food, particularly in meat and fish, the water vapour will leave the surface of the product and cause local dehydration, curiously this dehydration is called 'freezer-burn'. Freezer-burn results in discolouration, texture changes and poorer eating quality. Surprisingly, many enzymes are active during frozen storage and can produce changes in colour, flavour and texture in the food. Enzymes must be inactivated by blanching to prevent this occurring. However, recent work has shown that blanching can only inactivate some enzymes for a limited period and they are able to reactivate themselves during frozen storage.

When storing foods in a factory cold-store or a domestic freezer care must be taken to avoid tainting. Some foods, even when frozen, give out odours which are picked up by other foods, thus tainting them. It is estimated that only about 15% of all causes of tainting of frozen foods are discovered. The reason probably for this is that the taint may produce some off-flavour which might be described only as mousey, cat-like, mushroom-like and so on! Highly spiced foods, meats, fish, and many fruit products give off odours, whereas fats, particularly butter, meat and egg products, readily pick up odours. Good housekeeping is therefore essential when storing frozen food to prevent odour transfer.

Review

1. Chilled storage

– food kept above – 1°C and below 8°C, ideally below 4°C
– reduces spoilage by micro-organisms and enzymes
– inhibits the growth of thermophilic organisms
    (grow at 50–70°C), most mesophilic organisms (grow at 20–40°C)
– some mesophiles can grow at 5°C, all psychrophilic organisms (0–10°C) can
    grow in chill stores and spoil food
– some enzymes similarly can grow at chill temperatures
– chilled foods deteriorate gradually
– chilled foods must be handled carefully and hygienically
– chilled foods for sale must be labelled with either 'sell by' or 'best before' dates
– chilled stores may be more efficient if gas added (CA stores), particularly
    about 10% carbon dioxide

2. Frozen storage

– products must be frozen quickly and maintained at the same temperature
    without fluctuations
– best storage at −29°C/−22°F
– temperature fluctuations can cause loss of weight due to sublimation of ice
    and a number of chemical changes, such as rancidity in fat
– 'freezer-burn' is surface dehydration of the food caused by sublimation of ice
– frozen foods can cause tainting problems
– some foods give out odours and some readily accept odours

3.2 Cleaning of food raw materials

If a particular crop is harvested by hand the picker is usually very careful to avoid anything other than the particular food material. Other parts of the plant such as twigs, leaves and flowers are avoided. Similarly the picker will ensure that the product is not contaminated with stones, earth, string, weeds and obviously not insects, excreta or small animals. Modern mechanised picking machines and combined-harvesters have altered this, and contamination of food by the above means is quite common. In addition the machines can lose nuts and bolts, and drop grease or oil on to the food. Modern agricultural techniques make use of insecticides, herbicides and many fertilizers, all of which can leave residues on the food. Food raw materials must, therefore, be thoroughly cleaned before being processed further. Like so many preparatory processes in a food factory, cleaning is a *separation process* as the contamination has to be separated from the food. This separation can be achieved with or without water, but generally several stages are necessary to ensure that a product is thoroughly clean.

Cleaning *without* using *water* usually takes the form of sieving or screening any contamination from the product. Screens of different sizes can be used to separate large and small particles from the food material. This method of cleaning is cheap and fairly efficient, but with very dusty products there is a risk of recontamination from dust in the air. Like most continuous processes, continuous screening is more efficient and less labour intensive. In Figure 3.2 a continuous drum screen is shown in a simplified form.

Figure 3.2 Continuous drum screen

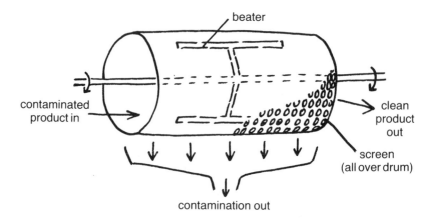

Root vegetables can be partially cleaned by rotating brushes which remove dried soil. *Aspiration* or winnowing is a development of the ancient process of throwing chaff and grain into the wind to separate them after threshing. The process depends on a strong upward air-current and the fact that the product and contaminants will have different buoyancies in this air stream.

Modern harvesting equipment has increased the occurrence of pieces of metal contaminating food materials. Should this metal reach the consumer in a prepared food product the manufacturer would be subject to a heavy fine. Magnetic material can be removed by powerful magnets, but aluminium and stainless steel, commonly found in food processing equipment, are non-magnetic. Metal detectors must be employed, therefore, to look for non-magnetic metals in the food.

Figure 3.3 Aspiration

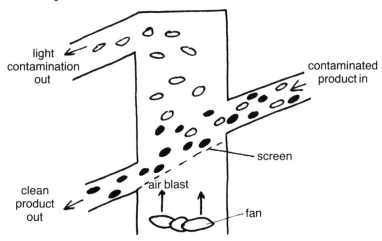

When *water* is used for cleaning it obviously must be free from contamination itself and should be free from bacteria and chemical contamination. Unfortunately in some parts of the world this is not the case and food products may become infected with typhoid, cholera and other diseases, and sometimes with poisonous chemical residues.

Heavily soiled vegetables can be *soaked* in water as a preliminary to further cleaning. The efficiency of soaking can be improved by agitation of the water. Spray washing is one of the most widely used methods of wet cleaning. A small volume of water at high pressure is used as the force of the spray cleans the food and it is not just a washing process with water running over the product.

Figure 3.4 Spray washing

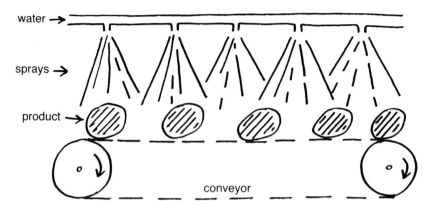

More complicated washing systems are used in some factories, for example, *flotation washing*. This system depends on a difference in buoyancy in water between the food and the contaminants. The food passes through a number of weirs and is forced under the water by slowly rotating paddles.

Figure 3.5 Flotation washing

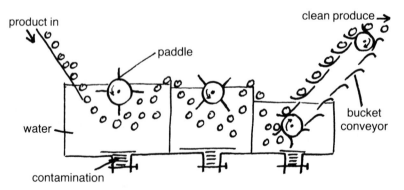

Generally a food material cannot be cleaned by just one method and it is necessary to employ perhaps three different methods to ensure that the product is properly cleaned. This is the first process involved in preparing a convenience food.

3.3 Sorting and grading

Like cleaning, sorting and grading are separation processes and sometimes similar equipment may be used, for example screens are used to sort by size. There is some confusion in the food industry between the terms, sorting and grading. *Sorting* should be reserved for the process of separating foods into categories of a single physical property, such as size, shape, colour or weight. *Grading* is quality separation and a number of factors may have to be assessed before a food can be graded into a certain category of quality.

3.3.1 Sorting

If individual food products are required to be processed mechanically then the job is made much easier if the food items are all of the same size. Filleting of fish can be performed mechanically, but fish must be of the same size, otherwise only a small fillet will be obtained from one fish and a large fillet from another including undesirable parts of the gut. Vegetables and fruit particularly can be sold in a supermarket in small trays and stretched-wrapped. Items which have been sorted into similar categories of size or weight are much easier to pack in such containers.

In processing involving heat penetration, such as blanching and canning, or in processes involving heat loss, such as chilling and freezing, uniformity of size is important. Large food items will be underprocessed and small items may be overprocessed.

Many fresh food items sold in supermarkets, such as meat cuts, are automatically sorted by *weight,* packaged and labelled. Small vegetables such as peas and beans, eggs, fruits and nuts are often sorted by weight.

A number of food raw materials are sorted according to *size* and a range of screens can be employed to do this. Round items are best suited to this type of process and they must be able to withstand quite rough handling. Fruits are often sorted by size, but care must be exercised to avoid bruising.

The natural *colour* variation of food products is normally acceptable to the consumer; however discolourisation due to rots, bruising and bad handling is not acceptable. Sometimes growing conditions and bad weather produce products with a lot of discolouration. Sorting by colour was carried out extensively using trained operatives in a tedious, labour-intensive manner. Unfortunately, this still cannot be avoided for some products. However, the process can be carried out mechanically employing a complex electronic system (sold under the trademark of Gunson Sortex) using photoelectric cells which compare the colour of the product with a standard colour background. Any of the products not matching the background are rejected, usually by a blast of compressed air. The method has been used successfully for many products, including dried peas, rice, coffee beans, grain, potatoes and onions.

3.3.2 Grading

Grading is separation into different categories according to the overall quality of the food. The term 'quality' is somewhat subjective as what is good eating quality in a product for one person may be poor for another. Take apples, for example; some people would consider an apple as good quality if it is firm and slightly acidic, whereas others consider a good eating apple to be soft, juicy and sweet. A good quality fruit for jam-making must be rich in pectin, but a fruit for juice production must be low in pectin, soft and juicy. In this case quality is dependent on the final end-use of the product. Normally the quality of a food cannot be judged by examining one aspect only. A number of factors have to be considered simultaneously to assess the food's quality, and obviously it is virtually impossible to mechanise this operation.

Work carried out at the Torry Fish Research Station, Aberdeen, has shown that in fish there is a change in the properties of the skin as the fish deteriorates. These changes in the skin cause a change in its dielectric properties which can be measured using a special meter called the 'Torrymeter' (Registered Trade mark). The scale of the instrument is such that a high reading equates to high quality fish and a low reading to older or spoiling fish. This instrument can be used by personnel after a little training, and is therefore a valuable grading tool.

First quality food products must be free from: damage; contamination; and undesirable parts of the raw material; and must be within a specified range of shape, size and colour. Texture also is an important factor in determining quality in such products as some fruits, vegetables, cakes and potato crisps.

Review

1. Cleaning

- can be with or without water
- contamination of food with stones, leaves, twigs and animal parts has increased due to mechanisation of harvesting
- use of chemicals in growing crops has increased chemical contamination
  cleaning is a *separation* process

- screening used extensively to separate soil and stones from food
- very dusty process but cheap and efficient
- aspiration using air blast to remove lighter particles from food
- water must be hygienic for wet cleaning techniques such as soaking, spraying and flotation washing

2. Sorting

- one characteristic considered such as size, weight or colour
- sorted foods better for mechanised operations, essential where heat transfer processes are used
- sorted foods easier to handle and pack into supermarket containers

3. Grading

- quality separation
- a number of factors must be considered simultaneously such as shape, colour, size, freedom from damage and contamination
- rarely will one attribute indicate quality
- dielectric properties of skin of fish can be used as indication of quality – measured by 'Torrymeter'

3.4 Size reduction

If you look at many packeted, tinned and frozen foods you will notice the products are not in the original size and shape but have been reduced or modified into a different shape. Obviously it is impossible to can a whole pineapple; make fish fingers from a whole fish; or bottle a fruit and call it juice. Size reduction plays an important role in the preparation of convenience foods and in preparing raw materials for other processes such as juice extraction.

We have seen in the processing of wheat into flour that the grain is *crushed* between grooved and then smooth rollers. Crushing is only used as a preliminary stage to further size reduction, but *grinding processes* are much more common. Special grinding machines include hammer, disc, pin and ball mills.

The *hammer mill* (Figure 3.6) is useful in shattering fragile material into small fragments. It is probable that this size reduction occurs in two stages. Firstly, the product fractures along existing fissures and defects, and then secondly new fissures are formed, followed by fracture along these fissures.

Disc mills use rotating discs with studs to break the product and *pin mills* use rotating plates with a large number of short metal rods or pins

Figure 3.6 Hammer mill

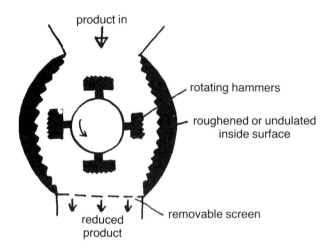

Figure 3.6 Hammer mill

Ball mills are tumbling mills consisting of a drum containing balls or rods which crush the product as the drum rotates. (Figure 3.7).

Figure 3.7 Ball mill

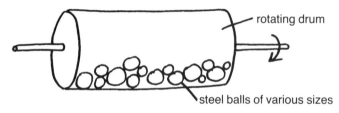

These types of mills rely on the fragility of the product and are totally ineffective for any fibrous, elastic or viscous product.

Fibrous materials have to be cut and cannot be reduced in size by crushing or impact. Sharp knife edges or saws are employed. Rotary cutting knives are often used for slicing meat products, fruit and some vegetables. Often slices are further reduced into cubes by the process of *dicing* in which cross-knives are used. Diced meat and vegetables are used in soups, dried products and some prepared complete meals.

Sometimes a food material is *shredded* as a preliminary operation to dehydration, where the large surface area aids the rate of water loss. Lower grade fruit is often *pulped,* by high speed paddles, and is then used in jam-making or frozen for future use.

3.5 Mixing

If you look at the list of ingredients on the side of a packet of dried vegetable soup you will see that there may be twelve or more different ingredients. If you open the packet you will notice large, medium and small particles of dried vegetables and a lot of powder. To blend these ingredients into a uniform mix is extremely difficult. The closer the particles are in size the easier the mixing process and the more difficult it becomes as particle sizes vary. Large variations in particle sizes can result in 'demixing' during a blending operation, so after a period of time the product may be less uniform than it was earlier in the process. When small quantities of one component have to be blended uniformly into large quantities of other components, some pre-mixing of the smaller component will be necessary in part of the larger component.

Liquids obviously do not have these problems of particle size and, therefore, mix easily when they are miscible. Some liquids separate out after mixing, eg oils and water and therefore *emulsification* is necessary (see Section 1.3.1.5). *Homogenisation* goes further than emulsification and is a mixing process combined with size reduction, as the dispersed liquid droplets are reduced in size in the crude emulsion and are mixed uniformly.

3.5.1 Mixing equipment

Mixing machines are very diverse and some are very specialized, for example dough mixers. Some mixers have more widespread uses and the mixing parts or *elements* can be interchanged, such as those on a kitchen mixer. *Liquid mixers* are designed to stir a liquid, but not in a regular flow pattern. When stirring a liquid in a bowl with a spoon it is normal to reverse the stirring action every few minutes in order to cause

Figure 3.8 Turbulence in mixing liquids

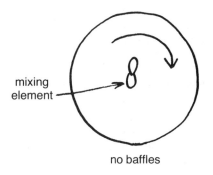

mixing element

no baffles

with baffles

turbulence and thus better mixing. This idea is practical in large-scale mixing, but baffles have to be used to obtain the turbulence, as shown in Figure 3.8.

The high speed propeller mixer is probably the most commonly used mixer for liquids, and is mounted off-centre in the mixing vessel to obtain turbulence (Figure 3.9).

Figure 3.9 Propeller mixer

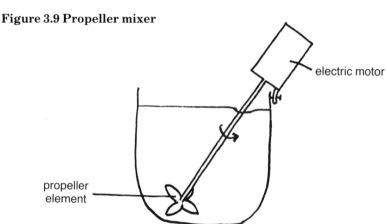

electric motor

propeller element

Powder and *particle* mixers rely on displacing parts of the mix in relation to other parts by some form of agitation. *Tumbler mixers* are used commonly for this purpose and revolve rapidly to tumble the mix. These mixers come in many shapes from cubes to double cones.

Figure 3.10 Tumbler mixer

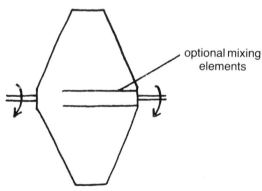

optional mixing elements

The *ribbon blender* is often used for dry mixes such as dried soups, instant desserts and bakery products. The mixer consists of a trough in which rotates a shaft with two helical screws, one screw being left-handed and the other right-handed. As the mixer operates one screw takes the mix in one direction while the other screw tries to bring the mix back again. Thus, thorough mixing is achieved.

Doughs and pastes are very difficult to mix because of their high viscosity, and mixers must be powerful with mixing elements capable of reaching all the 'dead-spots' of the mixing vessel. This can be seen by observing the action of a dough-hook attachment for a kitchen mixer. 'Z'-blade mixers are used for pastes in the food industry and consist of heavy mixing elements in the form of a 'Z' which twist in and out of the path of each other.

3.5.2 Homogenisation

During homogenisation two liquids are mixed and the particle size of one liquid, usually a fat, is reduced and then dispersed in the other liquid. Often an emulsifier is also needed to ensure long-term stability of the emulsion which is produced. In the case of homogenised milk (Section 2.1.1.2) the emulsifying system is naturally occurring.

To break down the liquid droplets the crude emulsion is forced through a narrow opening at high velocity. A *pressure homogeniser* is used for this and consists of a homogenising valve, or often two valves, and a high pressure pump (Figure 3.11).

Figure 3.11 Pressure homogeniser

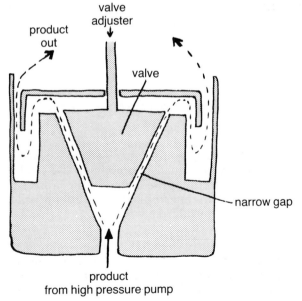

valve adjuster

product out

valve

narrow gap

product from high pressure pump

This type of homogeniser cannot be used for more viscous materials. The *colloid mill* can be used to homogenise more viscous products, such as mayonnaise. A rough or corrugated rotor revolves and the material passes between this rotor and the side of the mill through a very narrow gap. Homogenisation is thus achieved by this 'grinding action' in the limited space between the rotor and the side of the mill.

Review

1. Size reduction

- important in a number of processes as original material may be too large or wrong shape
- crushing – preliminary to other size-reduction processes
- grinding processes common, eg use of hammer, disc, pin and ball mills
- mills rely on fragility of product and are of no use for fibrous material
- sharp knife edges are required to reduce fibrous material, eg slicing, dicing and also shredding and pulping may be used for some products

2. Mixing

- the closer the particles are in size the easier the mixing process. Large particles are difficult to mix with small particles
- liquids easier to mix
- emulsification is necessary for liquids which are immiscible
- homogenisation involves reduction of the droplet size of one liquid
- when mixing liquids, turbulence should be encouraged
- powder and particle mixers rely on displacing parts of the mix in relation to other parts by some form of agitation
- tumbler mixer or ribbon mixers are common
- doughs and pastes require powerful mixers with heavy mixing elements
- pressure homogeniser used for liquids, colloid mill for homogenising pastes

3.6 Filtration and centrifugation

3.6.1 Filtration

Filtration in the food industry involves a very wide range of applications in separating liquids from solids. Sometimes the solid is retained for further processing, sometimes the liquid. In the process of *clarification,* filtration is carried out to remove small quantities of solid from a liquid, eg finely suspended material from wines.

Some terms used in filtration can be confusing. The *filtrate* is the liquid which passes through the filter and the filter membrane itself is referred to as the *filter medium.* Separated solids accumulate on the filter medium and these solids can be referred to as the *filter cake* or filter residue.

Filter media include fabrics, such as cotton, silk and nylon, paper, sand, charcoal, porous carbon and porcelain. The type used often depends on a particular application. The holes in the medium will vary in size in some materials, but can readily be blocked by solid particles during filtration. For this reason *filter aids* are often used. Filter aids consist of large particles of unreactive material, such as paper pulp and Kieselguhr, which form a lattice-like structure on the medium. The liquid can freely run through the lattice as the holes remain clear in the medium.

Figure 3.12 Action of a filter aid

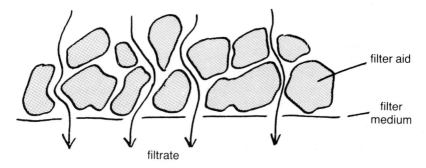

Figure 3.13 Section of a plate-and-frame press

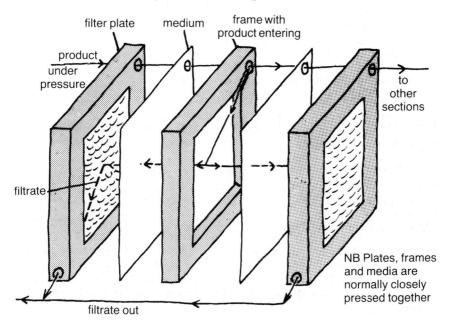

A very common filter used in many food industries is the *plate-and-frame press* (Figure 3.13). Special plates covered with filter medium alternate with hollow frames in a special rack. The plates, media and frames are squeezed together by an adjustable screw. The liquid to be filtered enters the frame and the solids build up on the medium while the pressure applied to the filter forces the liquid through the medium on to the plate. This is repeated in a number of sections in the filter.

As an alternative to applying pressure to speed up the filtration process, a vacuum may be applied to help draw the filtrate through the medium. *Continuous rotary drum vacuum filters* are commonly used as they are continuous and cheap to operate.

Figure 3.14 Rotary drum vacuum filter

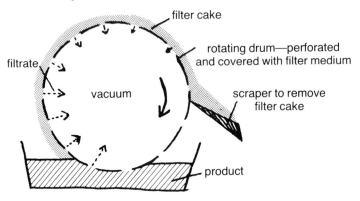

3.6.2 Centrifugation

Instead of removing solid particles from a liquid by filtration, they may be removed in many cases by centrifugation. We have already seen one application in the separation of cream from milk in Section 2.1.2. Centrifuges are used in purifying oils, clarification of beer, separation of yeast and in sugar refining.

3.7 Blanching

Blanching is a necessary preliminary operation in the preparation of vegetables and some fruit for canning, freezing or drying. There are a number of purposes for blanching. *Enzymes* must be inactivated by blanching prior to further heat processing otherwise they may cause flavour, texture and colour changes in the product. During canning, for example, until the temperature reaches a high enough level to inactivate the enzymes they will be extremely active.

It is, therefore, desirable to inactivate them by blanching before canning. In frozen products many enzymes show some activity, although often only slight in some products. This activity is enough to cause a loss in quality of a stored frozen product. (See Section 1.4.2.2.) Animal products are not blanched as enzymes are necessary to produce the desired flavour and sometimes texture. Fruits are rarely blanched since losses of volatile flavour compounds may be more critical than changes produced by subsequent enzyme activity.

An equally important aspect of blanching is the *shrinkage* it produces in a product, usually by the expulsion of trapped air or gases and the loss of some water. Mushrooms illustrate this problem well, as they contain much trapped air. If they are canned without blanching, the air is expelled in the can so that the can appears to be one third or less mushrooms, one third water and the rest air. Blanching reduces the mushrooms in size and expels most of the air. The filled mushrooms in the can still show some shrinkage during heat processing, but considerably less than previously. Cooking of mushrooms, of course, causes considerable contraction, so canned mushrooms should be compared with the normal cooked product.

Blanching has a number of side effects, some of which are beneficial, some deleterious. Blanching will help clean the product and will reduce the bacterial population. However, if the product is left wet and warm bacteria will multiply rapidly and may exceed their original numbers before further processing. In the case of some products, texture and colour are improved by blanching. Undesirable changes include losses in heat-labile nutrients, particularly vitamins. In the case of water blanching, water-soluble vitamins may be lost by leaching.

In blanching, foods are heated rapidly to a certain temperature, normally near 100°C/212°F, for the required time, then cooled quickly, or processed further without delay.

3.7.1 Blanching methods

The food material is dipped into *hot* or *boiling* water for a short period, usually 30 seconds to 3 or 4 minutes. In continuous blanchers the material passes through the water on a moving belt or rotating screw. The combination of time of blanching and the temperature is dependent on the size of the food material. The *peroxidase test* is used to determine the efficiency of blanching, see Section 1.4.2.2. Water blanching can cause high losses of soluble material in some products. In other cases additives can be placed in the blanching water to improve the product; these have included sulphites, citric acid and ascorbic acid (vitamin C).

Steam blanching uses saturated steam in a closed vessel through which the food is conveyed, usually by a rotating screw. Steam blanching offers advantages over the previous method in reducing losses of soluble material, but additives cannot be used.

Other methods of blanching have been investigated and *microwave* blanching offers a possible alternative method.

Products frozen in a domestic freezer are often used within, perhaps, three months. The likelihood of any changes being caused by enzymes is fairly remote. Many home-grown vegetables are frozen whole, eg cauliflowers, and it is impossible to blanch them satisfactorily when this large. The commercial frozen food company perhaps takes up to a year to prepare, store and distribute their products. Enzyme changes can be considerable in this period. Hence the need to blanch products for commercial frozen storage.

Review

1. Filtration

– separation of solid from liquid
– clarification to remove small quantities of solid from a liquid
– filtrate liquid passing through filter medium, solids remaining referred to as filter cake
– filter media include fabrics, paper, sand, porous carbon and porcelain
– filter aids, large inert particles, speed up filtration
– plate and frame press common pressure filter
– continuous rotary drum filter – works by vacuum to draw filtrate through medium
– centrifugation alternative to filtration

2. Blanching

– purpose to inactivate enzymes as they can damage product during canning, freezing and drying
– also blanching causes shrinkage by expelling air and some water
– blanching also cleans product, may improve texture and colour
– efficiency checked by peroxidase test
– water-blanching can cause loss of soluble materials, eg water-soluble vitamins
– additives, eg ascorbic acid, may be added to blanching water
– steam blanching does not cause losses due to leaching
– microwave blanching new development

Practical exercises:
Preparation of food raw materials

1. Frozen storage

Store samples of fish fillets in a freezer, both unpacked and packed well in polythene bags. Look for freezer-burn.

2. Cleaning

Use a sample of well-soiled carrots or potatoes.
(a) brush, with a small stiff brush, individual vegetables and note the inefficiency and difficulty of the process.
(b) soak vegetables in cold water and note the time most soil is removed. Agitate the water and note improvement in cleaning.
(c) Attach a rubber tube or nozzle to a cold tap, squeeze and spray the vegetables with a blast of cold water. Rotate the vegetables whilst spraying and note the time necessary to obtain a clean product.

3. Sorting and grading

Examine a range of convenience foods and note evidence of sorting and grading. Establish those criteria used for the grading of a particular product.

4. Mixing

Weigh equal amounts of sugar (say 200 g) and dried peas. Place in a large tin and seal. Shake for a few minutes. Remove a sample of the mixture and weigh. Separate out the peas and weigh. Repeat for varying lengths of time.

The best mix is when the peas and sugar are in equal quantities.

Repeat the experiment using a variety of mixing elements for a domestic mixer.

5. Filtration

Mix a sample of flour with water. Filter through normal filter paper. Note the filtering process becomes sluggish as the paper becomes blocked. Some starch will form a colloidal dispersion and pass through the filter. Repeat, adding a small amount of Kieselguhr to the mixture, note improved speed of filtration.

6. Blanching

Use a vegetable such as carrot or potato. Dice into $\frac{1}{2}$ cm^3 cubes.

(a) Water blanching – boil a saucepan of water. Place equal quantities of dried vegetable into three small muslin bags. Dip these bags containing the vegetables into the boiling water for 1, 2 and 3 minutes respectively. Cool immediately in cold water. Test for peroxidase activity as detailed below.

(b) Steam blanching – fill a pressure cooker with water just below the level of the trivet and heat to boiling. Repeat as for water blanching by placing a small bag of vegetables in the pressure cooker and then putting the lid on but not the pressure weight. Repeat for 1, 2 and 3 minute steaming.

(c) Microwave blanching – place a bag of vegetables in a microwave oven, and 'cook' for 1, 2 and 3 minutes. Test for peroxidase activity.

The peroxidase test

A spot of freshly prepared guaiacol solution (1%) is mixed with an equal volume of hydrogen peroxide (5 volume solution) and then placed on to the food. If peroxidase is still active in the vegetable a red/brown colour appears on the sample within one minute.

Section 4

Preservation Processes

If a food is not preserved by some method it will undergo spoilage by the action of micro-organisms and enzymes, and possibly by chemical reactions. Chemical reactions such as rancidity do not occur in all foods and are often a long-term spoilage. In most cases the preservation of a food is concerned with the destruction of spoilage organisms and enzymes, or in their inhibition over long periods of time. Heat processing is an example of the former in destroying organisms and enzymes by heat, and freezing is an example of the latter in that the organisms and enzymes can become active again on thawing.

Spoilage of foods by enzymes is often not so obvious, or as rapid, or such a potential health hazard as is spoilage caused by the growth of bacteria, fungi or other micro-organisms. A number of factors affect microbial growth and these can be made use of in the preservation of foods.

Exposing micro-organisms to *temperatures* above their maximum growth temperature will kill the cells of the organism, usually by denaturing the enzymes necessary for normal life functions. Some species produce very resistant spores which are difficult to destroy by heat. Temperatures below the minimum growth temperature of the organisms will retard growth (ie multiplication of cells) and freezing will halt growth completely. This is said to be a *bacteriostatic* process, as opposed to a process such as canning, which kills the bacteria, hence *bacteriocidal*.

Micro-organisms require *water* for growth; however some can grow at very low moisture levels. Removal of water from a food will inhibit the growth of organisms. Water can be removed by drying a food, but the water can also be made unavailable by freezing. Similarly, the addition of sugar and salt can deny micro-organisms their requirements for water by osmotic effects.

We have seen in Section 2 that there are a number of products which are preserved by lowering of *pH*. These include products where acid is added, eg pickles, or where acid is produced by fermentation, eg cheese, yoghurt and some pickles.

Some bacteria require *oxygen* to survive and similarly most fungi require oxygen. Some organisms, however, grow in the absence of oxygen, and some actually cannot grow in its presence. Organisms requiring oxygen are inhibited in a number of processes, such as canning, gas-packing and vacuum-packing by ensuring all oxygen is removed.

In Section 1.9 we saw that a range of chemical substances are permitted for use as *preservatives*. Antibiotics were once used for this purpose, but it was soon realised that bacteria could become immune to these substances, and could, on occasions, pass on this immunity to disease-causing organisms.

Micro-organisms can be destroyed or severely inhibited by a range of *radiation sources*, in much the same way as is any living organism. The use of irridation of foods as a means of preservation is the only new technology to emerge in food preservation in over a century.

Preservation of food sometimes relies on making use of more than one of these factors. Heat processing of canned foods which are acidic is normally less severe because of the lower pH.

Preserved foods must be packed into containers which prevent their recontamination. For this reason food packaging is included in this section.

4.1 Heat processing

Foods can be preserved by the application of heat in sufficient quantity to kill all micro-organisms and to inactivate all enzymes. There are, in fact, two levels of heat processing dependent on combinations of temperature and time. *Pasteurization* is heat processing designed to kill all pathogenic organisms, and in so doing to kill most spoilage organisms. This extends the storage life of the product a little but makes it bacteriologically safe (see Section 2.1.1.2 for the pasteurization of milk).

Sterilization is a much more severe heat process aimed to destroy all micro-organisms. Absolute sterility is never obtained as some bacterial spores may survive the process. *Commercial sterility* is the state achieved in most canning processes, and is heat processing designed to kill virtually all micro-organisms, and most spores, which would be capable of growing during storage.

The severity of the heat treatment of food to be preserved can be lessened if the food contains acid or osmotically active substances such as salt or sugar. Acidity is the most important factor affecting heat processing and its inhibiting effect on spoilage organisms starts at about pH 5·3. The presence of larger than usual amounts of fat, starch or sugar have been shown on occasions to protect micro-organisms at lower pH values than

5·3. The most important pH is 4·5, as below this pH the very dangerous organism *Clostridium botulinum* is inhibited. However, certain strains of the organism have survived lower pH values. The bacterium produces a very powerful toxin, if it survives heat processing of a canned food. The toxin has a mortality rate of about 70%. The spores of *Cl. botulinum* are very heat resistant and the canning industry is designed to ensure their elimination in canned foods.

Low acid foods have pH values of 4·5 or higher, and must be sterilized during heat processing. *Acid foods* have pH values of 4·5 down to 3·7 and a less severe pasteurization process will be sufficient to preserve them. At pH values below 3·7 only a few fungi can grow and in most cases the food can be considered a *high acid* food which may only require blanching to inactivate enzymes.

Foods must be heat processed in a container which will not be affected by heat and one which is sealed to prevent recontamination of the product. Traditionally foods were heat processed in cans and glass bottles, but now there is a rapid growth of pouches made of foil, plastics and special laminates capable of withstanding the process.

4.1.1 The heat process

If you look at the vast range of canned products you will notice that some contain liquid foods, eg soups; some solids suspended in liquids, eg canned peas; and some are completely solid, eg some canned meats. In the heat process the right quantity of heat must penetrate the can and its contents as far as the point in the can which is known as the '*cold point*'. In a *solid pack* such as canned meat the cold point is in the centre of the can and heat takes a long time to reach this point by the process of *conduction*. In the case of a *liquid pack,* such as soup, convection currents ensure rapid heat penetration into the product, and the cold point is moved slightly from the centre of the can.

If a special thermocouple (thermometer) is inserted at the cold point of a can the temperature history of the can may be recorded throughout the sterilization process.

As the food is heated the slope of the graph will be steeper if the food is a liquid pack until the holding-temperature is reached, the plateau part of the graph. Usually, the holding-temperature is 121°C/250°F, but a lower temperature is used for acid foods.

Some organisms, especially those of the genera *Bacillus* and *Clostridium,* produce heat resistant spores. Some of these spores may survive the heat process to germinate and grow in the food to spoil the food or produce toxic substances. *Cl. botulinum,* as mentioned above, produces a powerful toxin and is, therefore, used as an indicator organism in canning processes.

Figure 4.1 Heat penetration into cans

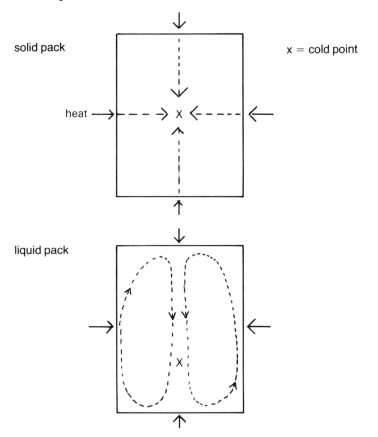

solid pack

x = cold point

heat

liquid pack

Figure 4.2 Temperature history in heat processing a canned food

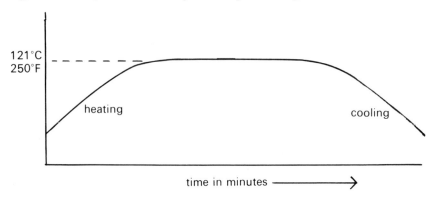

121°C
250°F

heating

cooling

time in minutes

Like other organisms and spores, the spores of *Cl. botulinum* are destroyed by heat at rates which depend on the temperature. At higher temperatures the rate of spore destruction is greater, and at lower temperatures spores are killed more slowly.

However, at any given temperature the spores are killed at different times, as some spores are more heat resistant than others. It can take a very long time at a certain temperature to destroy all spores and so a statistical approach is taken in evaluating the heat process and the heat resistance of organisms.

At a certain temperature, usually 121°C/250°F, after a period of time 90% of the spores will have been killed, and similarly after a further similar length of time, 90% of the remaining spores will be killed, ie 1000 spores will be reduced to 100, then 100 spores to 10. This period of time is known as the *decimal reduction time* or *D value*. At a higher temperature a shorter time or D value will be required to kill 90% of the spores. In Figure 4.2 the warming up, holding, and cooling periods of the heat process all contribute to the total sterilization process. Each combination of time and temperature can be calculated and added together to give the *F value* of the process. The F value denotes the sterilizing value of the process and is defined as the number of minutes at 121°C/250°F which will have a sterilizing effect equivalent to that of the process. F values will depend on the nature of the product and the size of the can involved; for example, beans in tomato sauce require F values from 4–6 minutes, while meat in gravy requires an F value of 12–15 minutes.

4.1.2 The canning process

The traditional canning process involves packing the food into a container, and then heating the container until its contents are 'commercially sterile'. Canning cannot improve the quality of a food raw material and at the very best can only maintain the harvested quality of the product. In general, however, there is a decrease in the quality of the product, which is often only slight using modern techniques. Unfortunately, in many parts of the world canned foods have a poor reputation, which is due in most cases to the use of low quality raw materials.

Canned foods are convenience foods and are intended to be ready to eat. Food raw materials must be cleaned effectively and free from all contaminants and inedible parts. The microbial load must be kept as low as possible prior to heat processing to minimise the risk of contamination with heat resistant spores. Canned foods usually require some size reduction, such as slicing or dicing.

Blanching must be carried out quickly after size reduction to avoid enzymic reactions, particularly browning reactions. The main purposes of blanching in canning are to expel air and shrink the product and to inactivate enzymes. After blanching the product is filled automatically either by weight or volume into the cans. Often with vegetables a 1–2% brine is added, which may contain other additives such as citric acid, colour and flavour compounds. Sometimes a sauce is added with a complex mixture of thickening agents, spices and herbs. The food is filled whilst hot and usually right to the top of the can.

The cans are passed to a special seamer where a vacuum is applied to draw out air from the can as the lid is sealed into place. A double seam is employed and involves rolling the 'hook' of the lid under the hook of the body of the can and then pressing it flat.

Figure 4.3 Double seam on can rim

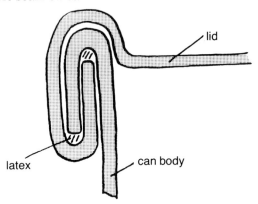

Usually some of the product spills out of the can and on to its outside during sealing, and this must be washed off. The cans are then heat processed in a retort (or autoclave) which is effectively a large pressure cooker.

The cans are usually placed in baskets which are lifted or wheeled on trolleys into the retort. The retort is closed and steam is allowed to enter. All air must be driven out of the retort by the steam, in much the same way as air is driven out of a pressure cooker. When steam is coming out freely from the vents of the retort all air is expelled. Any air-pockets between cans will cause under-exposure to the heat which might result in dangerous bacteria surviving in a can. The vents of the retort are closed, and the pressure rises in the retort, usually to 15 psi, which corresponds to a temperature of 121°C/250°F. After holding at this temperature for the required processing time the cans are cooled by spraying with cold water. Sudden application of cold water will cause the steam to condense too rapidly and the pressure developed in the can, which is equal to the outside pressure of steam, will push the can walls outwards and distort the shape of the can, straining its seams. For this reason, compressed air is applied to the retort as cooling begins to avoid this sudden pressure drop. Cans are often removed from the retort after partial cooling and are passed through a bath of chlorinated water to cool them to about 40°C/104°F. The remaining heat in the can dries off any

Figure 4.4 Canning mushrooms

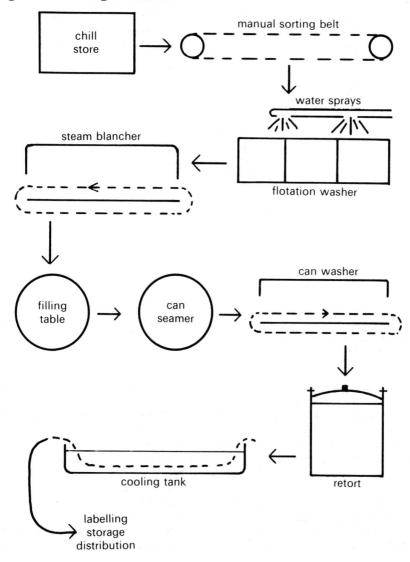

residual water on the surface of the can after cooling and this prevents any rusting. Cans are then labelled with special printed labels. In many cases now the cans are printed prior to the process and so labelling is obviated.

Cans which have been heat-processed must be handled with care especially if still wet. As the contents of the can cool a very strong vacuum builds up in the can. If there is any defect in the can or seam a small amount of water from the outside of the can may be drawn into the can. This water may be contaminated with food-poisoning organisms which can infect the food and cause an outbreak of food poisoning some time later. For this reason cooling water must be free from bacteria and should be chlorinated. A severe outbreak of typhoid was caused in Aberdeen when imported corned-beef from Argentina had been infected by contaminated cooling water. The water used in the cooling bath for the large cans of corned-beef was river water, from a river carrying typhoid organisms, and this water was not chlorinated.

Figure 4.4 is a typical flow-diagram of a canning line, in this case showing canning of mushrooms. Other lines will be similar but with variations particularly in size-reduction and mixing equipment for sauces and other additions.

Canning lines can readily be made continuous and very large throughputs can be obtained using continuous retorts. *Continuous* or *hydrostatic* retorts are extremely large and heads of water are used to maintain the pressure within the interior of the chamber of the retort.

4.1.3 Sterilizable pouches and glass

Tinplate for cans has become increasingly expensive, and although very thin plate is now used, the can is often more expensive than the food raw material. This has resulted in declining sales of canned foods, compared with frozen products which have shown a corresponding increase in sales. The flexible pouch, which is often a special laminate of foil and plastics, is a cheaper alternative to the can and has other advantages. The pouches are flat and therefore allow more rapid heat penetration during the sterilization process and processing times may be reduced by as much as 50%. However, pouches can be damaged easily and are often difficult to seal. Rates of production of pouches, therefore, do not compare with traditional canning lines at the moment. There is also a consumer resistance to pouches in the UK which is not shown throughout the rest of Europe or the USA. (Please see Section 4.5 for details of packaging.)

Glass is one of the oldest technologies and use of glass in food preservation dates back to the last century. Improvements in glass have been irregular but the most marked has been the production of lightweight but strong glass. Lightweight bottles, combined with wide-mouth closures, have resulted in a container which can rival the can for vegetables, fruits, beer and soft drinks.

4.1.4 Aseptic packaging

In Section 2.1.1.2 the problem of sterilizing milk was discussed with particular reference to the production of a 'cooked-milk flavour'. The UHT process was developed to sterilize the milk without causing the flavour changes. This technology can be applied to a range of products which are rapidly sterilized then packaged into sterilized containers.

Aseptic packaging can be defined as the filling of a commercially sterilized product into a container, previously sterilized, and applying a sterilized seal in a sterile environment to obtain a hermetic seal. The food is heated to a high temperature 150°C (300°F) for a short period of only a few minutes and then cooled. Direct injection of steam may be used for some products to obtain the same result. The sterilized product is filled into sterile cans, in a sterile environment, which are then hermetically sealed.

There are a number of plants operating this system for soups, ice-cream mixes and custards. Work is being carried out with the use of special micro-filters to remove bacteria and thus sterilize the product, where the product is particularly heat-sensitive.

4.1.5 Spoilage of canned foods

In 1812 the firm Hall & Donkin (England) used tinplate to produce cans of roast veal. In 1818 some 46,000 lb of these canned foods were purchased by the Admiralty. A few remaining cans were opened in 1938 and were found to be in perfect condition. However, on occasions, cans which were usually too large, were under-processed and food-poisoning outbreaks resulted. It was not until 1922 that the current style of can was generally accepted.

Under-processing of cans now is a rarity, but trouble can occur due to the presence of very heat resistant spores. *Thermophilic gas spoilage* is caused by the growth of *Clostridium thermosaccharolyticum* which has particularly heat resistant spores. The organism produces large amounts of hydrogen which causes the can to bulge and even burst. Some bacteria can produce souring of products which are normally low in acid but without the obvious production of gas. *Bacillus stearothermophilus* is a common causative organism of this problem which is only detected when the can is opened and the contents tested. *Clostridium nigrificans* produces hydrogen sulphide which can blacken the contents of an under-processed can of non-acid foods. Sometimes, however, bulging of cans and some discolouration can be caused by natural acids in the food, for example, canned tomatoes often show some swelling.

As we saw in Section 4.1.2 poor seams and can defects can lead to *'leaker spoilage'*. Unlike under-processed foods, even bacteria which show no heat resistance may contaminate the product. Bacteria entering the can may spoil the product and produce gas, causing the can to swell. Sometimes, harmful bacteria may grow in the product without any gas production. *Salmonella typhi* (typhoid) normally produce gas in cans and so they can be discarded as obviously being infected. However, in the Aberdeen typhoid outbreak the nitrite of the corned-beef inhibited the gas production and the organisms grew and eventually caused the outbreak of typhoid.

Damaged cans usually have experienced strain on their seams and organisms may have entered. Damaged cans should be avoided, as should any showing swelling.

Review

1. Preservation processes in general
    – foods spoil by action of micro-organisms, enzymes and chemicals
    – number of factors affect microbial growth:
        (1) temperature
        (2) water
        (3) oxygen
        (4) pH
        (5) preservatives
        (6) radiations

2. Heat processing
    – foods preserved by application of sufficient heat to kill micro-organisms and enzymes
    – *but* pasteurization kills all pathogens and some spoilage organisms
    – sterilization more severe, kills all cells and most spores
    – 'commercial sterility' achieved in canning as some spores may survive
    – severity of heat treatment lessened in the presence of acid or osmotically active substances such as sugar
    – *Clostridium botulinum* – dangerous food-poisoning organism – inhibited usually below pH 4·5
    – low acid foods pH 4·5 or higher require sterilization
    – acid foods pH 4·5 down to 3·7 require pasteurization
    – high acid foods pH below 3·7 require blanching only

3. The heat process
– 'cold point', part of can which is slowest to heat up
– in solid packs heat transfer is by conduction
– in liquid packs heat transfer is by convection
– combinations of temperature and time required to kill organisms
– higher temperature = shorter time
– Decimal reduction time or D value is time to reduce spores to $1/10$th of original level
– F value is the number of minutes at 121°C/250°F which will have a sterilizing effect equivalent to that of the process

4. The canning process
– pack food into container, seal, then heat process until contents are 'commercially sterile'
– canning process must include:
  (1) cleaning of raw material
  (2) blanching
  (3) filling – addition of brine or sauce
  (4) sealing – double seam formed
  (5) sterilization – in retorts
  (6) cooling
  (7) labelling
– retorts operate usually at 121°C/250°F
– all air must be expelled prior to processing to avoid pockets of air where under-processing will occur
– cooling water must be chlorinated to avoid 'leaker-spoilage'
– sterilizable pouches (laminates of foil and plastics) and toughened glass alternatives for cans
– aseptic packaging – product sterilized by high temperature, short time process then filled into sterilized containers under sterile conditions
– cans may spoil if under-processed, as spores will survive to produce gas, souring or sulphide discolouration
– 'leaker-spoilage' may occur if seams damaged or can is dented – any organism can be involved

Practical exercises: *Heat processing*

1. Heat penetration
Take a can of tomato soup and a can of meat (dog food will suffice), and remove the lids. Suspend a thermometer in the centre of the can contents, using a retort stand. Place the cans in a water-bath and boil. Record the rise in temperature of the product against time, and plot a curve of heat penetration, with temperature against time in minutes.

2. Effect of metals on texture of heat processed vegetables
Make up solutions (0·5 litre) of sodium hydrogen carbonate (bicarbonate) and calcium chloride in strengths of 1%, 2%, 5% and 10%. Boil samples of vegetables

such as peas, carrots or diced potatoes for 1 minute in each of these solutions, and in water. Cover samples of each in a beaker and keep in a refrigerator for several days. Note any change in texture as the firming effect of calcium may take several days. Sodium samples should be noticeably softer.

3. Can examination

Examine the outside of the can and note body seams, reinforcement rings and labelling. Open the can and look for presence of lacquer on the inside of the body and the lid. Cut into the side of the can and cut across the double-seam with a suitable saw. Examine the seam with a hand-lens.

4.2 Freezing

In Section 3.1 the storage of foods under chilled and frozen conditions was discussed. If foods to be frozen are placed in a cold store, a very poor quality product is produced. However, the quality of most frozen foods can, by careful control of the freezing process, be equal to that of the original 'fresh' food material.

Freezing preserves foods by two principles. The very low temperatures involved inhibit microbial growth, often causing the death of some organisms, and also retard the action of enzymes or chemicals. The production of ice during freezing causes water to be withdrawn from the food, and this dehydration effect also prevents microbial growth.

As we have seen in Section 1, most foods contain a large amount of water, which is normally in the form of solutions of sugars, salts, acids and other substances. These substances lower the freezing point of water and so food normally freezes at a temperature below 0°C/32°F. During a freezing process most foods begin to freeze at about -1°C. It takes some time for all the water in a food to freeze, particularly in large items. As water is withdrawn to make ice crystals, the dissolved substances become more and more concentrated. This increase in concentration of the solutions in the food depresses the freezing point further. Often the temperature must fall to about -5°C before most of the water is frozen. Even at this temperature concentrated solutions will remain unfrozen and may contain up to 20% of the original water. When no more ice crystals are formed, the food cools rapidly, usually to about -30°C/ -22°F. This last part of the cycle (see Figure 4.5) is called 'tempering'.

The *'thermal arrest period'* is the time the food takes to pass through the part of the cycle where water is frozen into ice (-1°C to -5°C). To produce high quality frozen foods it is necessary for this period to be as short as possible. The thermal arrest period will depend on the size of the product. For peas the period will be, perhaps, one second, but a side of beef may take 36 hours to pass through the period.

Figure 4.5 Freezing cycle of a food

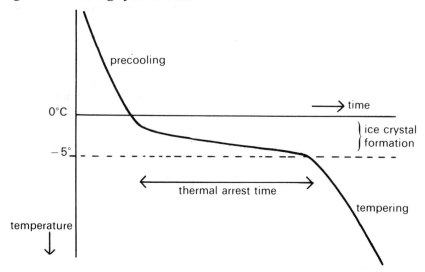

When plant materials are frozen, the solutions outside the cells freeze first of all as they are less concentrated than solutions inside the cell. Ice crystals begin to form in this way between the cells and steadily increase in size (Figure 4.6). Slow freezing rates allow water to be withdrawn from the cells to form large ice crystals outside the cells.

Figure 4.6 Ice crystal formation in plant products

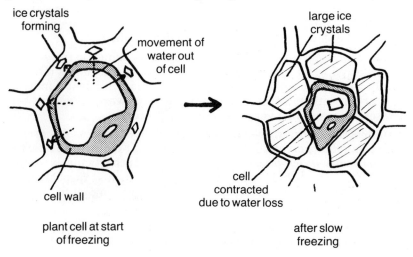

The cells become dehydrated by this means, and on thawing they will be found to have lost all their turgor pressure and will be in a collapsed state. The large ice crystals outside the cells may push into the cells and cause some damage, but on thawing they will produce a lot of water, which is not within the cells, giving a very watery texture to the product. Slow freezing causes the concentration of substances in the cells because of this water loss, and these solutions may damage the proteins or other cellular constituents. If the temperature is lowered rapidly the water outside the cells is again first to freeze, but is rapidly followed by freezing of the water within the cells. Only small ice crystals are produced which do not cause concentration of solutions within the cells and the accompanying damage. On thawing the water produced from melting ice is still within the cells and the texture of the product is closer to that of the original product.

Fish and meat products do not have such an obvious cellular structure, but have different types of tissue. Water can be withdrawn in a similar manner from the tissues, and, on thawing, produces a pool of water around the product. This water is known as 'drip' and is noticeable in fish which has been frozen slowly.

The highest quality foods, therefore, must be frozen quickly and must have their final temperature as low as is practicable. The term 'quick-freezing' is a vague term which is often used. The term means that the thermal arrest period is as short as possible and the product is reduced to its final storage temperature quickly. For fish, which is quick-frozen, the thermal arrest period should be less than two hours and the final recommended temperature should be $-30°C/-22°F$.

Deep-freezing has been defined by the International Institute of Refrigeration as a process whereby the average temperature of the product is reduced to $-17·8°C/0°F$, then kept at this temperature or lower. This definition does not take into account the rate of freezing which can be slow, producing a lower quality product.

An old term, *'sharp-freezing',* is misleading as domestic freezers have been referred to as 'sharp-freezers'. The rate of freezing is often very slow. In a domestic freezer this rate of freezing is sufficient if individual items are frozen on a tray, then packaged afterwards. Packing before freezing will result in a very slow freezing process.

In commercial freezing operations, high quality products are produced by adopting this individual freezing method, the products are referred to as *individually-quick-frozen* (IQF). The method is particularly useful for vegetables, such as peas, and berry-type fruits, such as raspberries. The IQF process is continuous, followed by packing into bags for supermarket sale.

Re-freezing of frozen products should always be avoided, as quality will fall each time the product is frozen. Bacteria, which often are only inhibited by freezing, will begin to multiply in the thawed product and may reach dangerous levels. On re-freezing, then thawing again, levels of organisms may be high enough to cause food-poisoning.

4.2.1 The refrigeration cycle

The most common system employed in freezers and other refrigeration systems is that of *vapour compression*. Refrigeration is generally understood to be the method of causing heat to flow in a direction which is not natural, for example, from a cold substance to a hotter one. This reversal of flow of heat is obtained using a *refrigerant,* which is a substance having a high latent heat capacity on changing state from gas to liquid and back again.

Exchange of heat occurs at two surfaces, an evaporator and a condenser. These surfaces are joined and connected to a compressor and expansion valve, as shown in Figure 4.7.

Figure 4.7 Vapour compression refrigeration cycle

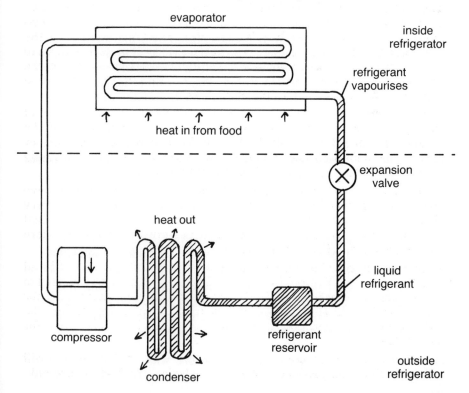

Liquid refrigerant takes up heat in the evaporator from the food to be frozen. The heat causes the refrigerant to evaporate and the vapour formed is pumped away from the evaporator by the compressor, where it is compressed, and then condensed back into liquid refrigerant in the condenser. The liquid passes round the circuit to the expansion valve which releases it into the evaporator to complete another cycle.

Ammonia has been used extensively as a refrigerant, but the freon group of organic compounds is now used widely, particularly in smaller refrigeration systems.

4.2.2 Freezing methods

Immersion-freezing was one of the first methods generally employed for freezing foods. It is a slow method with little control. Brines were prepared from salt and ice; as more and more ice was added the temperature fell to a suitable level for freezing. The food was then immersed in this brine for several hours. Modern methods have used refrigerants which have been sprayed onto the food. The method is little used as the following methods offer greater advantages.

Plate-freezing has played a major role in making available a wide range of frozen produce, and has only been replaced by other methods in the last few years for some products. The food is prepared in the normal way and is packed into a flat container, usually a cardboard based container, often with a polythene or wax lining. The container is placed between flat, hollow, refrigerated metal plates and the plates are adjusted to press tightly against the pack. The method depends on contact between the pack and the freezing plate and any air gaps will slow heat transfer from the product, thus greatly increasing the freezing time.

The plates may be horizontal (Figure 4.8) or vertical. The horizontal type was commonly used for many types of food products, but the vertical type is used mainly for bulk items. Vertical plate-freezers are often used on board trawlers for freezing fish fillets into blocks, which can be sawn up into fish fingers at a later stage.

The main disadvantage of plate-freezers is that they cannot easily freeze irregular shaped food materials. The *blast-freezer* is ideally suited for this type of operation.

Generally a blast-freezer is a large cabinet in which a fan has been introduced to move the air over the product. Stationary air acts as an insulator, but moving air readily takes up heat and loses it again quickly. The effect can readily be experienced by blowing on the back of your hand, especially when wet. Ideally the air should have a temperature of $-25°C/-10°F$ or lower and should move with a velocity of about 400 m/min. The blast-freezer (Figure 4.9) generally is a batch system and, as such, has disadvantages, being only suitable for a low throughput of product. Continuous systems have been developed and have much larger throughputs.

Figure 4.8 Horizontal plate-freezer

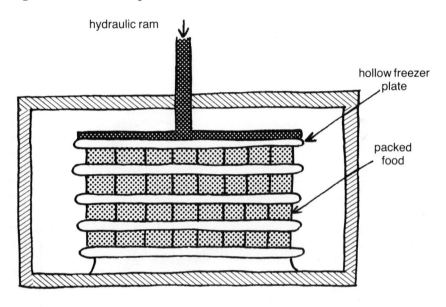

Figure 4.9 Blast-freezer

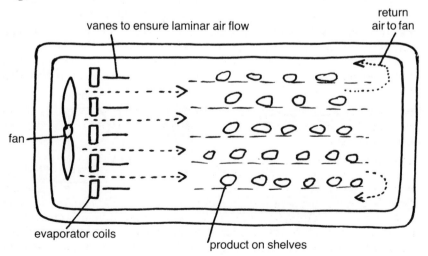

Fluidised-bed freezers are blast-freezers with a vertical air blast of sufficient velocity to 'fluidise' the product in the air stream. To fluidise a product, such as peas, beans, chipped potatoes and soft fruit, the product must be suspended on jets of refrigerated air. Freezing takes between four and ten minutes and such freezers are continuous, often with throughputs of 10 tons/hour (Figure 4.10).

Figure 4.10 Fluidised-bed freezer

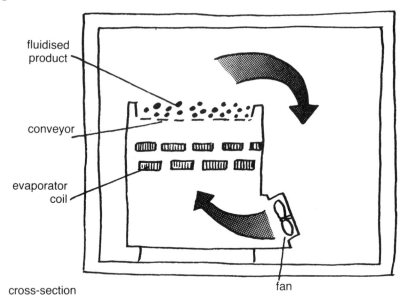

fluidised product

conveyor

evaporator coil

cross-section

fan

Figure 4.11 Liquid nitrogen freezer

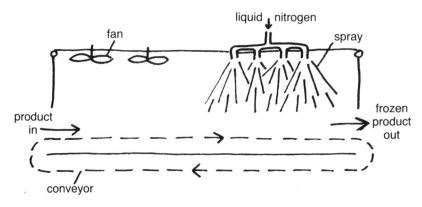

liquid nitrogen

fan

spray

product in

frozen product out

conveyor

224

Cryogenic-freezers make use of very cold liquified gases, such as nitrogen and carbon dioxide. Freezing is so rapid for some products that they suffer thermal shock because of the rapid contraction caused by the sudden lowering of temperature. Liquid nitrogen (at $-196°C/-320°F$) is sprayed on to the food on a conveyor-belt in a tunnel (Figure 4.11). Nitrogen gas is removed by fans to the entrance of the tunnel to pre-chill foods before the actual freezing stage, to reduce thermal shock. The method is good for soft fruits and products such as prawns, but not larger items, as too much nitrogen will be used.

Liquid CO_2-freezers work at a higher temperature $(-78°C/-108°F)$. However, the system is more economical than the nitrogen system and the gas can be recovered and re-used. Because liquid CO_2 cannot exist at atmospheric pressure, it is stored in bulk under pressure and reduced temperature. As soon as the pressure is released the CO_2 converts to a fine powder of carbon dioxide snow, and an equal amount of CO_2 gas. The CO_2 snow has a surface temperature of $-78°C/-108°F$ and converts directly to CO_2 gas on contact with the food, thus achieving rapid freezing rates. The carbon dioxide is also a bacteriostat and helps preserve the food product. Frozen storage systems have been developed, using liquid CO_2, for use on airliners.

4.2.3 The cold chain

All frozen food has to travel down the 'cold chain', from the freezer to the bulk cold store, via refrigerated vehicles to wholesalers, then retailers, and finally to the home. As we saw in Section 3.1 fluctuating temperatures during the storage of frozen products can lead to considerable loss in quality. Freezing and storage in the factory should be at $-30°C/-22°F$ for best results and long storage potential. Some heat gain must occur when the product is transferred from the cold store to the transport vehicle. The vehicle itself, although it is refrigerated with a mechanical refrigeration or liquid nitrogen system, can only maintain the temperature of the load. The vehicle's refrigeration system can only cope with heat coming through the insulation, doors and floor, and cannot reduce the temperature of the product.

The wholesaler's cold room will not be at such a low temperature as that of the factory and may be only at $-20°C/-8°F$, or occasionally $-25°C/-13°F$. Smaller vehicles, possibly relying on insulation and without a refrigeration system, may transfer the product to the retailer where the cold room will be at $-18°C/0°F$ or higher.

The retailer usually makes use of open-top display cabinets for frozen products. These cabinets have caused numerous problems at this stage of the cold chain. Over-filling will mean that packs on the top will be at too high a temperature, and the use of strip-lighting may warm packs close to the top and back of the cabinet. The cabinet should be, nevertheless, at about −15°C/5°F and never warmer than −12°C/11°F.

When a pack of frozen food is bought, it is perhaps carried round the supermarket, put in a bag and then taken home during a time period possibly of hours. Some products can thaw out completely during this time. Frozen packs used to be wrapped in newspaper, a moderate insulator, but now a token plastic bag is used.

If the food is used immediately the quality loss resulting from the journey home will not matter; however, if the food is placed in a freezer problems may result. A partially thawed food will re-freeze at a much slower rate, producing large ice crystals and a product of lower quality.

In 1964 a *star marking* system was introduced for frozen foods and also for domestic freezers and refrigerators. The star markings are given in Figure 4.12.

Figure 4.12 Star marking for recommended storage times for frozen foods

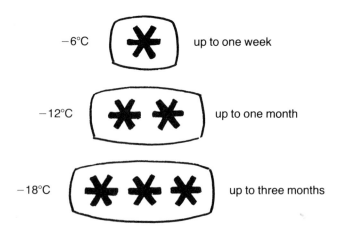

−6°C up to one week

−12°C up to one month

−18°C up to three months

The use of similar markings on the food packs and appliances give easy cross-reference.

In 1973 a further symbol (Figure 4.13) was added to indicate a freezer capable of freezing food as well as storing frozen food.

Figure 4.13 Symbol indicating food-freezing capability

When this symbol is used on a freezer the manufacturer must state the weight of food which can be frozen in each 24 hours. This weight factor is called the *rated freezing capacity* of the appliance. The three stars, as before, indicate that frozen foods can be stored up to 3 months.

The cold chain is represented graphically in Figure 4.14. Any method of handling and storage which reduces the slope of the graph will improve the final quality of the product.

Figure 4.14 The cold chain

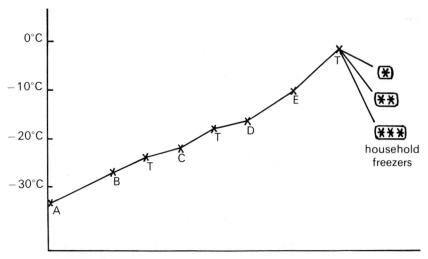

A: factory freezers
B: factory cold-store
C: distribution/wholesalers cold-store
D: retail cold store
E: display cabinet
T: transport

/iew

_ reezing

– preserves by: (1) low temperature
(2) withdrawing water in the form of ice
– dissolved substances lower freezing point below 0°C/32°F, food begins to freeze at about −1°C
– thermal arrest period – the time the food takes to pass through the part of the freezing cycle where water is frozen into ice
– slow freezing leads to low quality produce as water is withdrawn from the cells
– large ice crystals form outside the cells; on thawing form 'drip'
– rapid freezing produces small ice crystals, a higher quality product and less drip
– 'quick-freezing' – thermal arrest time as short as possible and product reduced to final temperature quickly
– 'deep-freezing' – average temperature of the product is reduced to −17·8°/0°F then kept at this temperature or lower
– individually-quick-frozen (IQF)
  – individual freezing, eg of peas.

2. Refrigeration cycle

– main system involves vapour compression
– refrigerant takes up heat in evaporator from the food – vapour formed pumped away
– compressed then condensed back to liquid
– liquid passes to pressure release valve
– enters evaporator and cycle is repeated

3. Freezing methods

(1) Immersion freezing – use brines or refrigerants
(2) Plate freezing – freezing by contact with refrigerated plates, can be horizontal or vertical
(3) Blast freezing – cold air blown on to food of any shape
  – temperature and speed of air control rate of freezing
  – fluidised bed-freezers use vertical blast of air
(4) Cryogenic freezing – use of liquified gases at very low temperature, usually nitrogen or CO_2

4. The cold chain

– frozen food has to travel down cold chain:
  freezer → bulk cold store → refrigerated
  vehicles → wholesalers → retailers → home
– heat gain at all points leading to quality loss
– display cabinets can be over-filled and strip lighting can warm product
– product may thaw after purchase if journey home is too long
– star marking used to indicate storage life of a product and storage potential of an appliance

Practical exercises: *Freezing*

1. Freezing of meat

Take a sample of minced meat and form into a burger. Insert a thermometer into the centre of the burger, or, if available, use a thermocouple and digital thermometer. Place in a freezer, repeat with a sample in the quick-freeze section and in the freezer compartment of a refrigerator. Every 5 minutes observe the temperature and plot the freezing curve. (The process may take some time as the rate of freezing will be slow.)

2. Observations on frozen foods

Observe the levels of filling products into retailer display units. Note the time taken to buy a frozen food and take it home. Look for any evidence of thawing out. Place in a domestic freezer and note time taken to refreeze (this will not always be possible with well packed foods). Construct a diagram representing the cold chain for the product.

3. Chilling of fish

Take a whole fish and take its temperature. Pack crushed ice around the fish and record the temperature and time. Plot the cooling curve for the fish. Make a 1% salt solution and add some ice to bring the temperature down to 0°C. Immerse the fish and repeat the cooling process. Compare the ice and chilled water (corresponding to refrigerated sea-water) for their ability to chill fish.

4.3 Dehydration

Water is required by micro-organisms for them to maintain a normal population growth. Removal of water by dehydration does not kill the microbes but just stops their growth, in a comparable manner to freezing. If the moisture level of the dried product rises due to pick-up of water, from the atmosphere or any other source, micro-organisms will again resume their activity and reproduce, ultimately to spoil the food.

Some micro-organisms require a high level of moisture to be able to grow, whereas some, such as osmophilic yeasts, can grow at very low moisture levels. The amount of available water in a food is described in terms of water activity (a_w). Moisture levels are compared with pure water which has an a_w of 1·0. In Table 4.1 are listed ranges of water activities which are required by certain micro-organisms. Drying is carried out to reduce moisture levels to a very low figure, but as we have seen in Section 2 some products are preserved by salt and sugar which reduce the a_w by osmotic effects. Typical examples of lowered water activity are found in jams, salted foods and sugar products.

Table 4.1   Water activity ranges

Water activity	Organisms which grow in this range
0·90 and above	Most bacteria
0·85–0·90	Yeasts
0·80–0·85	Moulds
0·75	Halophilic bacteria (spoil salted products)
0·60–0·75	Osmophilic yeasts (spoil products high in sugar)

Some water in food is *bound* and cannot be easily removed by dehydration. However, nearly all water normally present in a food can be removed by one of a number of drying processes. The loss of water results in a significant reduction in the weight and bulk of a product, as well as preserving it. This enables a greater weight of actual food nutrients to be transported in a given volume of a container or vehicle. Famine relief measures clearly indicate this useful aspect of dehydration, as, for example, large amounts of dried milk powder are sent to disaster areas and not bulk containers of liquid milk.

Any drying method will therefore cause shrinkage of a food material, but this is minimal with freeze-drying techniques. Rapid drying systems cause the outer edges and corners of the food piece to become dried out and rigid and thus fix the shape of the dried food piece early in the process. Water is removed from the centre of the food to produce a light honeycombed product which readily rehydrates when added to water. Slow drying allows the product to shrink further and produce a dense dried food which is difficult to rehydrate.

As a food dries, water moves to the surface where it is evaporated. Water is always in the form of a solution in foods and the substances dissolved in the water cannot escape by evaporation and so accumulate at the surface of the product. While there is still water in the food some of these substances will diffuse back into the centre. However, in many foods there is an accumulation of soluble solids at the surface of the dried product. These solids may prevent complete dehydration and slow rehydration of the product when it is used.

Sometimes the accumulation of salts and sugars at the surface, combined with heat, causes a skin to be formed which is known as *case-hardening*. This has been known to occur when drying fruit, meat and some fish products. Case-hardening prevents complete dehydration and inhibits rehydration. To minimise the problem some products are pierced with a pin-hole prior to drying.

Many drying processes involve heat which causes the destruction of some vitamins, particularly C and B_1. In dried foods containing fats the loss of moisture and the concentration of salts has been shown to accelerate oxidative rancidity reactions. There is always a change in shape, colour and eating quality after drying, but the process of freeze-drying has been found to minimise these changes.

4.3.1 Drying processes

As we have seen in Sections 4.1 and 4.2, certain pre-treatments are necessary before dehydration can be carried out on a particular food. Food materials dry more rapidly if they are reduced in size and this produces a greater surface area for water loss. Many products, such as vegetables and meat, are diced prior to drying. Enzymes will become very active in these initial stages of a drying process and so they should be inactivated by blanching. Some products are also treated with a variety of substances, for example, sulphite may be added to the product to minimise enzymic browning reactions, or non-enzymic reactions such as the Maillard reaction.

Sun-drying is an ancient process which is still one of the main methods of drying a food in some countries. In hotter countries a vast range of products is dried by the sun, particularly fruit, fish and meat products. The method is often slow and is uncontrolled. There are often considerable problems with insect infestation, particularly of fish. The product can shrink excessively and can take a considerable time to rehydrate. In Britain, peas (marrowfats) used to be dried in the field but produced a variable product which had to be soaked, sometimes all night, to rehydrate completely.

Warm air driers are very common in producing a wide range of products, particularly cheap dried vegetables. A simple system is the *kiln drier*

Figure 4.15 Kiln drier

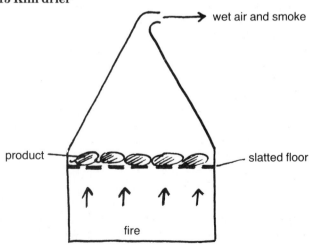

231

which is often a two-storeyed building, with a furnace on the ground floor producing warm air which rises through a slatted floor to dry the product (Figure 4.15). (This method has been used for hops for many years.)

Tunnel driers are a very common type of hot air drier. Some produce rapid drying of the product with little shrinkage, but do not obtain very low final moisture levels (the concurrent type Figure 4.16). Other types slowly dry the product, making it shrink more and of poorer rehydration ability, but of very low final moisture levels (the counter-current type).

Figure 4.16 Tunnel drier

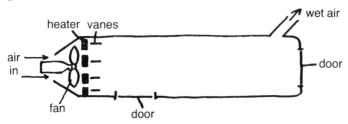

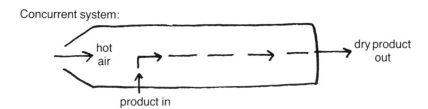

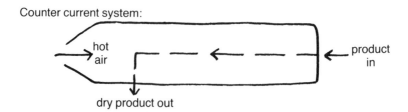

These tunnel driers are comparable with blast-freezers and in a similar manner it is possible to make use of the air-flow to 'fluidise' the product in a *fluidised bed drier* (Figure 4.17). A fluidised bed drier can be a continuous drier producing a product of low shrinkage and good rehydration characteristics. Its use is limited to products which are small enough to fluidise by the upward blast of hot air in the drier.

Figure 4.17 Fluidised bed drier

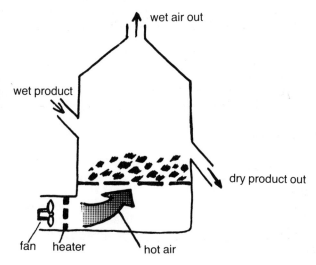

None of the driers described so far can be used for liquid foods. A common method, which was used for some time, was the *roller* or *drum drier*. This method was used for milk products, baby foods and breakfast cereals. In most cases it has been superseded by the spray drier. A large stainless-steel drum is heated internally by steam and this drum rotates slowly. A paste or concentrated mixture of the product is picked up by the drum and is dried as the drum rotates, to be scraped off by knives on the opposite side, Figure 4.18. Sometimes two drums are used which rotate

Figure 4.18 Roller or drum drier

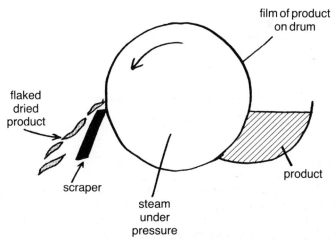

towards each other. High product temperatures can be achieved which produce a sterile product, but one which may be heat-damaged, causing browning reactions, loss of vitamins and decrease in protein quality or solubility.

The *spray drier* is used extensively to dry a wide range of liquid foods, particularly milk, egg products and products such as desserts. A fine spray of the product is produced by an atomizer which is usually built into the top of a large conical shaped chamber. On entering the chamber the spray of food is met by a blast of hot air, which dries the food into a fine powder within seconds. As the process is so fast bacteria may survive, and so it is necessary to heat-process the liquid before drying. Milk powder produced by this method is a very fine powder which does not wet or disperse well in water. To improve the wetability, dispersability and ultimately the solubility of the product, it is often re-wetted! This re-wetting causes the powder particles to clump together to make sponge-like structures which are dried by warm air in a fluidised-bed drier. These sponge-like structures absorb water rapidly and disperse and dissolve in it quickly. As an alternative to re-wetting the powder may be only partially dried initially.

All the drying processes described above have involved heating to some extent with the accompanying changes produced in the product. *Freeze-drying* produces a very high quality product as it involves little or no heating. As a consequence of this there is little shrinkage in the product; less flavour changes; no case-hardening; and good rehydration characteristics. The product is very friable, however, and its honeycombed structure readily crumbles if not carefully handled.

The process of freeze-drying involves freezing the product by a normal freezing method and then subjecting the frozen product to a very strong vacuum. Instead of the ice melting it sublimes to leave the product in a dry state. However, as the process is very slow, and therefore expensive, the process of *accelerated-freeze-drying* (AFD) was developed. The

Figure 4.19 The principle of accelerated freeze drying

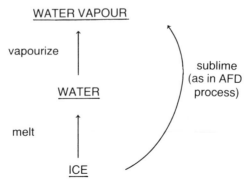

product is subjected to a strong vacuum and a small amount of heat by conduction, radiation heating or microwaves. The sublimation process is, therefore, accelerated and costs are reduced. Continuous AFD units are now in operation.

Very high quality products, particularly dried fruit and vegetables, meat and coffee have been made by this method.

The economics of any food factory necessitate the operation of large, continuous processes. Drying has made greater advances than many other processes because it has been actively involved in the development of such convenience foods as dehydrated vegetables, soup powders, instant coffee, desserts and instant mashed potato. The products are convenient to carry; they store easily; and they can be used often to make a meal in a few minutes involving minimal preparation.

Review

1. Dehydration

- amount of available water in a food, described in terms of water activity (a_w), where pure water has an a_w of $1·0$
- lower a_w levels are achieved by use of sugar, salt and dehydration
- some water is bound in foods and difficult to remove by drying
- quick drying causes less shrinkage than slow drying and produces a product which rehydrates more easily
- soluble solids move to surface during drying – some diffuse back to centre
- case-hardening – skin on surface caused by soluble solids, heat and changes in proteins. Slows dehydration and inhibits rehydration in water

2. Drying processes

- pre-treatment necessary: size reduction, eg dicing, blanching, sulphiting,
- sun-drying – slow, uncontrolled, but popular in hot countries
- warm air-drying – kiln drier, simple method
  - tunnel driers popular for vegetables
  - concurrent drier – less shrinkage, good rehydration characteristics in product, but final moisture levels higher
  - countercurrent drier – more shrinkage, poorer rehydration but low moisture levels
- for liquids – roller drier – high product temperatures, bacteria destroyed but loss in quality and browning may result
  - spray drier – very popular, rapid method producing fine powder – bacteria may survive, heat pretreatment necessary
- freeze-drying – very high quality product
  - freeze product then subject it to high vacuum to cause sublimation of ice to leave product free of water – AFD – some heat applied to accelerate process

Practical exercises: *Dehydration*

1. Drying vegetables

Dice a number of carrots evenly into about ½ cm^3 cubes. Weigh a sample of about 40 g. Place on a tray or, if available, on a fine mesh wire-rack. Place in a fan assisted oven at the lowest temperature setting, usually about 70°C. Remove the sample every 10 minutes and re-weigh. Plot a curve of weight against time until the product is dry. Repeat at a number of other temperatures. Repeat using diced carrots which have been blanched for 2 minutes in boiling water. Note the physical characteristic of each dried sample.

2. Rehydration of dried vegetables

Add samples of dried vegetables to water and gently heat. Note the time and extent of rehydration. Compare with commercial samples dried by known methods such as tunnel drying and AFD.

4.4 Irradiation of foods

We have discussed, so far, only one method which has a lethal effect on micro-organisms, and that is sterilization by heat. Irradiation is the only other technology which has a lethal effect and can be compared with heat processing in a number of respects.

Irradiation is the only completely new technology to be applied to foods in over a century, but it is probably the most studied of all technologies. At first the technology was thought to have great potential as a means of reducing microbial spoilage and insect damage; improving organoleptic properties of foods; obviating the need for some chemical additives; and extending the life of unrefrigerated foods. All other technologies were thought to be on the verge of redundancy. However, a number of serious setbacks occurred in the 1960's and approval in the few applications, then in operation, was withdrawn. Research has continued throughout the world, particularly with regard to the wholesomeness of irradiated foods.

4.4.1 Wholesomeness of irradiated foods

All organisms can be damaged or changed in some way by a number of radiation sources. Spoilage and food poisoning organisms can be destroyed by radiation at a certain level, but sometimes this level is much higher than would kill a human being. Ionization of a food component can occur when the food is irradiated, leading to the formation of reactive molecules and free radicals (see Section 1.3.4 for the action of free radicals in fat rancidity).

Other side-reactions might include the production of off-flavours and the destruction of vitamins.

The irradiation process is unlikely to induce radioactivity in foods, except, perhaps, for some trace elements. However, the chance of residual radioactivity in irradiated foods has caused considerable consumer resistance to the possible use of this technology on a large scale. Any doubts about this residual activity have now been removed by extensive animal feeding studies (International Project in the Field of Food Irradiation) and by chemical tests. These investigations have produced no evidence of adverse effects as a result of irradiation. Breakdown products of food components, ie water, carbohydrates, fat and protein in irradiated foods are the same regardless of the food, and therefore toxicity testing of all foods irradiated with less than 10 kilogray (kGy) is not necessary*. Most of these breakdown products are also found in other processed foods, particularly those which have been heat processed. These breakdown products are known to be safe and there is no evidence of toxicity.

4.4.2 Legislation

Irradiated foods are permitted in over 30 countries and are quite widely accepted in a number of these. In 1986 a report was published by the Advisory Committee on Irradiated and Novel Foods which concluded that the irradiation of foods proved to have no significant disadvantages or risk to health. From January 1 1991 the use of irradiation was legalised in the UK for some vegetables, spices and poultry products.

4.4.3 Irradiation processes

Irradiation plant is expensive, particularly because of the heavy screening to protect operatives. To operate economically plant must be used continuously. As only a few commodities can be irradiated legally in some countries, economical operation is extremely difficult.

Two types of radiation are used: electrons produced in a linear accelerator from radioactive sources; and gamma rays from the radioactive decay of cobalt 60 and caesium 137 sources. The electron beams are restricted to foods of less than 5 cm in thickness because of their poor penetrability. Gamma rays, which unlike electron beams cannot be directed, are more common and cheaper to use.

*World Health Organisation, Technical Report Series 659 Geneva 1981.

In a continuous irradiation operation the food material should be exposed to the radiation source at various angles to minimise any penetration problems.

Radiation sterilization or *radappertization* is intended to produce a sterile product similar to a canned food. Doses of radiation have to be high, in excess of 10 kGy and often as high as 30 kGy. Unfortunately, high dosages of radiation on a number of products, particularly fruit and vegetables, have been found to cause changes in colour, texture and flavour. Most development in recent years has therefore been carried out at low dosage levels.

Radiation pasteurization or *radurization* gives a significant reduction in the number of spoilage organisms and destroys pathogenic organisms. Levels of radiation of the order of 2·5 kGy are used, but changes in the spoilage pattern may result. Very large quantities of potatoes have been irradiated in Japan since 1973 to inhibit sprouting (dose level 0·1 kGy). In Holland, chicken has been treated by irradiation since 1976. Also the Dutch have irradiated successfully smaller quantities of diced vegetables, shrimps, froglegs and spices.

In South Africa a range of products have been treated, usually at a dose level of 0·7 kGy. The range of products includes potatoes, chicken, mangoes, strawberries, bananas, onions, garlic and avocados.

Products, such as potatoes, other vegetables and fruit receive low dosage levels to inhibit sprouting or ripening. However, controlled atmospheric storage or refrigeration may still be necessary to ensure prolonged storage. Products which are sterilized by irradiation must be aseptically packaged, and, as with heat processed foods, recontamination should be made impossible.

Irradiation will only become widely accepted after the complete picture of all aspects is made public. It is a question of education to overcome all misapprehensions.

Review

1. Irradiation

- only completely new technology for over a century
- potential to reduce:
  - (1) microbial spoilage
  - (2) insect damage
  - (3) need for chemical additives, also to improve organoleptic and keeping qualities of unrefrigerated foods. Inhibition of sprouting and ripening in fruit and vegetables

2. Wholesomeness of irradiated foods

– extensive tests have shown there are no risks with irradiated foods
– in some products high dosage levels can cause flavour, colour and texture changes
– irradiated foods are permitted in over 30 countries
– irradiation in the UK permitted in the Food Safety Act from January 1 1991

3. Irradiation processes

– plant is expensive as it needs extensive screening for operatives
– two types of radiation:–
  (1) electrons produced from radioactive sources in a linear accelerator
  (2) gamma rays emitted from cobalt 60 or caesium 137
– electrons have poor penetration, gamma rays cheaper, with better penetration but not directional
– radiation sterilization (radappertization)
  – high dose levels 10 kGy (kilogray) to 30 kGy
  – causes colour, flavour and texture changes in some products
– radiation pasteurization (radurization) similar to pasteurization in killing pathogens and some spoilage organisms, dose about 2·5 kGy
– very low levels (0·1 kGy) found to be useful in inhibiting potato sprouting – largest use in Japan

4.5 Packaging

From the earliest times some form of packaging has been necessary for food. The prime functions of packaging have remained the same throughout time – to protect, contain and identify the product. Prevention of recontamination of a preserved product is the prime purpose of many examples of packaging, particularly in canning, and is the reason for including packaging in the preservation section of the book.

Natural materials such as leather, pottery, wood, gourds and baskets were used for thousands of years. Processed materials came into use perhaps three hundred years ago with the adaptation of glass, paper and metal. It is only in the last forty years that synthetic materials, such as cellulose and polythene films, have been available. These materials have been adapted to rapid changes in the food industry, particularly to increasing rates of production demanding cheap but strong packaging materials.

The packaging of food must *protect* the food during storage, transport, sale and the journey to the home. The package must protect against *mechanical* forces, such as impact, vibration or compression. Very fragile foods, such as *AFD* fruits, need to be well protected, usually by packing into a can or foil laminate. Fibrous and tougher food products require less protection against mechanical force. A number of *climatic conditions* can affect foods. Some foods are susceptible to humidity, high temperatures, light and different gases. These facts must be considered when choosing a package material for a particular food. The packaging material must prevent *contamination* of a food by micro-organisms, insects, chemicals or soil. A package must be shaped to *contain* the food conveniently, but sometimes this is difficult. Problems are often overcome by other means, for example, special small cylindrical carrots are grown exclusively for canning. The normal tapered carrots of varying sizes are difficult to can.

To *identify* the product the package is labelled, often with considerable detail about the product, serving instructions, ingredient declarations, and average weight of product. Very tight regulations govern the labelling of foods, and even the name of the product is carefully regulated. For example, a can labelled 'Beef Curry' must contain at least 35% beef, whereas a product labelled 'Curry with beef' need only contain 15% beef. The label on a package is designed to be attractive and 'eye catching' by using different colours, shapes and designs. Often a photograph of the product is used, but this must not be misleading. To return to the example of beef curry, the label would show a picture of the curry served with rice. This would imply, at first sight, that the can included rice as well as the curry. The words 'serving suggestion' are often included under the photograph, and other articles not included in the package are often slightly out of focus.

4.5.1 Packaging materials

4.5.1.1. Food cans

Food cans (or tins) consist of a body with a fixed end, into which the food is filled, to be followed by sealing an identical end on the top of the can. Cans were rather heavy and consequently with rises in the cost of materials became more expensive than some of the products they contained. Weights of cans have now been reduced by over 50%, with thinner walls supported by strengthening ribs.

Traditional cans are made from tinplate which consists of thin low-carbon steel protected from corrosion by a thin layer of tin. New developments are taking place to eliminate the tin coating and replace it with special lacquers. The use of aluminium has increased enormously, particularly for drinks.

Electrolytic action can result in the dissolution of any exposed iron, and sometimes, in the presence of acid, hydrogen gas is evolved. *Lacquers* are often used to prevent any interaction between the package and the product. These lacquers are wax-like materials with specific functions.

During heat processing, sulphur-containing amino acids such as cysteine can be broken down to liberate sulphur compounds, particularly hydrogen sulphide. This hydrogen sulphide will react with the can and produce black iron and tin sulphides which will discolour the product. For this reason, when canning fish or meat products, a sulphur-resisting lacquer (epoxy-phenolic lacquer) is used on the tinplate and ends of the can. In some liquid packs, such as peas and soups, sulphur compounds can accumulate in any head-space in the can, giving an unpleasant smell when the can is opened. Sulphur-absorbing lacquer (oleo-resinous lacquer) is used with zinc oxide, which produces white zinc sulphide and is not noticeable. Acids will attack the tin-plate, and if lacquered inadequately gas production will result. Acid will concentrate its attack through any pin-holes in the lacquer and will produce a 'hydrogen-swell'. Recent lacquering techniques have lessened this problem.

Some food pigments, notably the anthocyanins, react with tinplate. The pigments produce a grey sludge in the can and the can must be totally lacquered to prevent it.

4.5.1.2 Paper and board

Paper and board are still the most used packaging materials. Raw materials include chemical woodpulp, which is mainly cellulose; mechanical woodpulp, which is cheaper and brittle; and recycled waste paper.

Board must be specially prepared for direct contact with food. *Solid white board* is commonly used and is produced from chemical woodpulp which is bleached. Other board, such as *chipboard*, is produced from re-cycled waste paper which has a natural grey colour. The board is used for outer containers which are not in contact with the food. Wax coating or polyethene may be applied, particularly for packaging frozen foods.

No major changes have occurred in many of these packaging materials for several years. However, a recent innovation is the 'ovenable' board used to make shallow trays for prepared or frozen foods which are cooked by microwaves. The carton board is coated with polypropene which can resist temperatures up to 140°C/284°F. Foil trays cannot be used in microwave ovens.

4.5.1.3 Plastics

A very wide range of plastics is used for food packaging either as flexible or rigid containers. These materials include 'polythene' (low density polyethene), ethene vinyl acetate (EVA), polyamide (nylon) and polyvinylidene chloride (PVDC). Each film can be made in a variety of densities and often with a range of properties.

The most common plastic packaging film for food is 'polythene' which is low density polyethene (polyethylene). It is a relatively cheap film, offering good properties which include: particularly good heat sealing, water vapour resistance, strength and low temperature resistance. It may be printed or coloured to improve its appearance. However, it is not a good oxygen barrier and cannot withstand temperatures above about 90°C/194°F. High density polythene will resist boiling temperatures and has been used for 'boil-in-the-bag' products. Linear low-density polythene is a new development produced by a low-pressure process. The film has improved performance over conventional polythene in having improved temperature resistance and greater strength, allowing thinner films to be used.

The poor oxygen barrier characteristics of polythene can be improved if it is laminated with another film, particularly polyamide (nylon) which is a very good barrier to oxygen. This is essential for vacuum packaging a number of foods which deteriorate in the presence of oxygen, for example, foods containing fat. A new development which may offer advantages for some products is the production of vacuum-metallised films in which a minute layer of aluminium is deposited in a vacuum chamber on to a plastic film base. This can be further laminated if necessary and can be made into retortable pouches. If better barrier properties are required the aluminium foil is laminated with paper and polythene. The paper carries the label and provides strength. The foil provides a very good barrier to gases, moisture and light, and the polythene provides a heat sealing facility with some protection for the foil.

PET (polyethylene terephthalate) bottles have made a significant impact on the soft-drinks industry, as they offer reduced container weight, large size, and little risk of breakage. The plastic is rigid enough, but is aided by gas pressure, to which it has excellent barrier properties.

Polystyrene in a number of forms is used for trays and insulated containers. Supermarket trays are made of expanded polystyrene which can be used to keep produce cool.

4.5.1.4 Cellulose films

Cellulose films are available in a range of densities, characteristics and uses. The films are coded by letters which have specific meanings, as shown in Table 4.2.

Table 4.2   Cellulose film codes

Letter(s)	Meaning
P	Permeable (not a moisture barrier)
M	Moisture proof (nitrocellulose coated on both sides)
DM	Demi-moisture proof (coated on one side)
QM	Quasi-moisture proof (slightly moisture proof)
MXXT	Very good moisture barrier (PVDC coated)
MXDT	Good moisture barrier (PVDC coated on one side)
B	Opaque
C	Coloured
S	Heat sealable

4.5.1.5 Glass

Glass has been used for food packaging for a considerable time. However, for high speed production in a food factory glass had severe limitations, because of the ease of breaking. New light-weight but tough glass containers have changed this. Wide-mouth closures and light-weight bottles are gaining popularity for beers, soft drinks and some jars are available for fruit and vegetables. A sleeve of expanded polystyrene, which may be printed, can be used to protect some types of bottles.

4.5.2 Packaging individual foods

The technologist needs to know the nature of the food to be packaged and the degree of protection required. A few examples of the problems involved are given below.

Fresh *fruit and vegetables* continue to respire and transpire in a package. The package must be permeable to gases, as oxygen is required for respiration. A somewhat decreased oxygen supply may prolong storage life. Continued transpiration, however, causes the accumulation of water droplets on the inside of the package, which is likely to facilitate mould growth. Packaging films are often perforated to release this moisture. Leafy vegetables are packed to control too much water loss, to prevent wilting.

Fresh *meat* requires oxygen in order to maintain its bright red colour. In the absence of oxygen, oxymyoglobin is lost and myoglobin, a dull red, is formed. A good moisture barrier is needed to prevent too much weight loss in the form of water vapour, but it must be a barrier which will allow the passage of gases, particularly oxygen. *Cured meats* do not require oxygen, which may damage their colour by causing the formation of brown metmyoglobin. A film impermeable to oxygen, and gas packaging, would be beneficial for these products.

Dehydrated foods must be kept dry and only a small amount of moisture can cause a loss in keeping quality. Many dried products are easily oxidized and so a barrier impermeable to moisture and oxygen should be used. Often dried foods are fragile and some protection is needed against mechanical change, by using a can or foil laminate.

Packaging materials for *frozen foods* must have low moisture permeability to prevent 'freezer burn', which is surface dehydration. Gaps between the food and the packaging material must be avoided, otherwise sublimation of the ice from the food may occur and this will refreeze next to the packaging material, producing 'frost', which is unsightly and also results in product weight loss. Low oxygen permeability is desirable to minimise flavour exchange and tainting problems in a cold store.

Figure 4.20 Bar code

4.5.3 Bar coding

This system of coding packaged articles was introduced in 1978, although it has been in operation in USA since 1973. Every product is assigned its own particular number which also indicates its price and description. The number appears on the pack in the form of bar symbols which can be read rapidly by electronic scanners at the supermarket checkout. The system ensures rapid 'checkout', which is also completely accurate and avoids assistant's errors. Movement of the bar symbol over the scanner automatically signals a computer which feeds back data, such as price, and this is automatically printed on the till ticket.

An example of the European Article Number (EAN) is given in Figure 4.20. This system uses a thirteen digit symbol.

Review

1. Packaging

- must protect against mechanical forces, climatic conditions, and contamination
- must contain the product – aided by plant breeding, eg for canning
- must identify the product: name, storage conditions, serving instructions, ingredients declaration, and average filled weight

2. Packaging materials

- cans – low-carbon-steel protected by thin layer of tin
  - now 50% reduced in weight
  - lacquer to protect against sulphur staining of product, colour changes, eg anthocyanins with iron, and acid attack
- paper and board – most common method of packaging
  - solid white board made from chemical woodpulp
  - chipboard made from recycled paper
- plastics – either flexible or rigid
  - most common 'polythene' ie low density polyethene
  - high density polyethene resists boiling temperatures
  - polythene – poor barrier to oxygen
  - 'nylon' (polyamide) good oxygen barrier
  - aluminium foil can be laminated with polythene
  - new plastic bottles – PET (polyethylene terephthalate), very light but strong.
  - cellulose films – 'cellophane' – wide range of types, denoted by letter codes eg P, permeable; M, moisture proof
- glass – new lightweight, but strong glass, becoming popular

3. Packaging individual foods

- need to know nature of food and degree of protection required
- fresh fruit and vegetables require oxygen, but must allow loss of some moisture
- fresh meat requires oxygen to form bright red oxymyoglobin
- cured meat needs protection against oxygen to prevent discolouration
- dehydrated foods protect against moisture, oxygen and mechanical damage
- frozen foods protect against moisture loss to prevent 'freezer-burn' and require low oxygen permeability to minimise tainting problems

Practical exercises: *Packaging film*

Try the following on samples of packaging films.

1. Water tap test – hold a piece of film under a running tap. Film which is not moisture-proof soon becomes saturated and limp.

2. Breath test – breathe on a sample of film; a moisture-proof film will cloud over.

3. Burn test – light with a match one end of a small, tight, roll of film. Observe the burning, smoke and odour produced. Does the material drip molten beads?
   cellulose films – burn sometimes slowly, do not melt or drip, eg P, MS & DMS
   polystyrene – burns with black smoke, and melts
   polyester – does not burn easily, but melts
   polyethene – burns slowly, melts and drips.

4. Tear test – tear a piece of film.
   cellulose are easy to tear
   polyester shows some tear resistance
   polyethene is difficult to tear.

Index

Page numbers in heavy type indicate the main references to a particular subject.